Road Atlas
for the
Total Solar Eclipse
of
2017

Black & White Edition

Fred Espenak

Edition 1.0
October 2015

Road Atlas for the Total Solar Eclipse of 2017 – Black & White Edition

Astropixels Publishing
P.O. Box 16197
Portal, AZ 85632

Astropixels Publishing Website: *astropixels.com/pubs*

This book may be ordered at: *astropixels.com/pubs/Atlas2017.html*

Printed in the United States of America

ISBN 978-1-941983-08-9

Astropixels Publication: AP009 (Version 1.0a)

First Edition

Front Cover: A map of the 2017 eclipse path in the vicinity of St. Louis illustrates the cartographic features found in the ***Road Atlas for the Total Solar Eclipse of 2017 – Black & White Edition***. Map copyright © 2015 by Fred Espenak. More about the eclipse can be found at:

eclipsewise.com/solar/SEnews/TSE2017/TSE2017.html

Table of Contents

The Sun's corona is only visible to the naked eye for a few minutes during a total eclipse of the Sun. This image was captured during the total solar eclipse of March 29, 2006 from Jalu, Libya.
© 2006 F. Espenak, www.MrEclipse.com

1.1 Introduction

On August 21, 2017, a total eclipse of the Sun will be visible from the continental United States for the first time since Feb. 26, 1979. Although a partial eclipse will be seen from all of North America, the total phase in which the Moon completely covers the Sun (known as *totality*) will only be seen from within the narrow path of the Moon's umbral shadow as it sweeps cross the USA.

It is only during *totality* that the Sun's faint and exquisitely beautiful outer atmosphere – the solar corona – is revealed to the naked eye, the landscape is plunged into an eerie twilight, and brighter stars and planets appear. The spectacle of a total solar eclipse is one of the most remarkable in nature. But to witness it you *must* be in the *path of totality*.

The path of the Moon's shadow begins in the northern Pacific and crosses the USA from west to east through parts of the following states: Oregon, Idaho, Wyoming, Nebraska, Kansas, Missouri, Illinois, Kentucky, Tennessee, Georgia, North Carolina, and South Carolina. The width of the *path of totality* ranges from 62 miles (99 km) along the Oregon coast to a maximum of 71 miles (115 km) in Kentucky.

The duration of totality also varies along the path being shortest in western Oregon (1 minute 59 seconds) and longest in southern Illinois (2 minutes 40.3 seconds). These durations are for the middle of the path along the *central line*. As you move away from the *central line* (either north or south), the duration of *totality* decreases. It happens very slowly at first but then drops rapidly to *zero* as you reach the northern or southern edges or limits of the *path of totality*.

This collection of maps was designed to assist eclipse observers in locating and reaching the central line of the eclipse in order to enjoy the longest possible totality.

1.2 Overview Maps of the Path of Totality

The first 3 maps give a broad overview of the total eclipse path through the western, central and eastern USA. The yellow lines running somewhat vertically through the eclipse path mark the position of mid-eclipse at 10-minute intervals from Oregon through South Carolina. The times are given in local time at each position along with the duration of totality on the central line, and the altitude of the Sun above the horizon at that instant.

1.3 Detailed Maps of the Path of Totality

A series of 37 maps covers the entire path of totality from Oregon through South Carolina. The *Table of Contents* (pages iii – iv) can be used to quickly navigate to the map of interest since it lists the state, local time and major city featured on each map.

The map scale is approximately 1:700,000, which corresponds to 1 inch ≈ 11 miles (1 cm ≈ 7 km).[1] This large scale shows both major and minor roads, towns and cities, rivers, lakes, parks, national forests, wilderness areas and mountain ranges. A 20 mile (30 km) reference scale appears at the bottom of each map.

The path of totality on each map is depicted as a lightly shaded region with the northern and southern limits clearly labeled. The total eclipse can be seen only inside this path (a partial eclipse is visible outside the path). The closer one gets to the central line, the longer the total eclipse lasts. Gray lines inside the path mark the duration of the total eclipse in 20 second steps. This makes it easy to estimate the duration of totality from any location in the eclipse path.

Note that each of the detailed maps has been rotated by small angle in order to keep the eclipse path horizontal. This allows the eclipse path to be seen in context with the regions immediately outside the path. A compass indicates the direction of north in one of the upper corners of each map.

The local time of mid-eclipse is marked by a series of lines crossing the eclipse path every 2 minutes. (Abbreviations for local times are: PDT = Pacific Daylight Time, MDT = Mountain Daylight Time, CDT + Central Daylight Time, and EDT = Eastern Daylight Time.) The eclipse circumstances on the central line are labeled with the local time of mid-eclipse, the duration of totality (minutes and seconds) and the altitude of the Sun.

Among the larger cities in the umbral path, there are 3 located on or just inside the northern limit (Bowling Green, SC; Lincoln, NE; St. Louis, MO), and another 4 on or just inside the southern limit (Charleston, SC; Kansas City, KS; Nashville, TN; North Platte, NE). As a consequence of this close proximity to the path limits, the duration of totality in these cities varies significantly depending on one's exact geographic location. In fact, some parts of Lincoln, Kansas City and St. Louis actually lie outside the path, in which case only a partial eclipse will be visible from those areas. On the other hand, Knoxville, TN, is completely outside the path so its citizens will witness a partial eclipse. Fortunately, Knoxville lies within 6 miles (10 km) of the northern limit, so it's just a short trip to get into the path of totality.

All maps were produced using Google Maps as the underlying map with overlying eclipse graphics generated using Javascript code. A web page is available to the user for examining any part of the 2017 eclipse path at a range of zoom magnifications. An added benefit of the web page is that it automatically calculates the local circumstances for any point the user chooses, with just the click of a mouse. For more information and to access the interactive 2017 eclipse path plotted on Google Maps, visit:

eclipsewise.com/solar/SEgmap/2001-2100/SE2017Aug21Tgmap.html

1.4 EclipseWise.com Web Site

For many years the NASA Eclipse Web Site was the leading Internet resource for predictions and information on eclipses of the Sun and Moon. The webmaster of the site (Fred Espenak) has now retired it as a location for new information but it remains an archival site for eclipse predictions prior to 2015. All future predictions for upcoming eclipses will be posted on the new web site *EclipseWise.com*.

---

[1] Because of the Mercator map projection, the actual scale on a given map can vary by up to 10% of this value.

EclipseWise.com has individual web pages, maps and diagrams for every solar and lunar eclipse from 2000 BCE to 3000 CE. This amounts to 11,898 solar eclipses and 12,064 lunar eclipses. Much of the design, layout and graphics were inspired by the recent publications *Thousand Year Canon of Solar Eclipses 1501 to 2500* and the *Thousand Year Canon of Lunar Eclipses 1501 to 2500*. (See: *astropixels.com/pubs)*

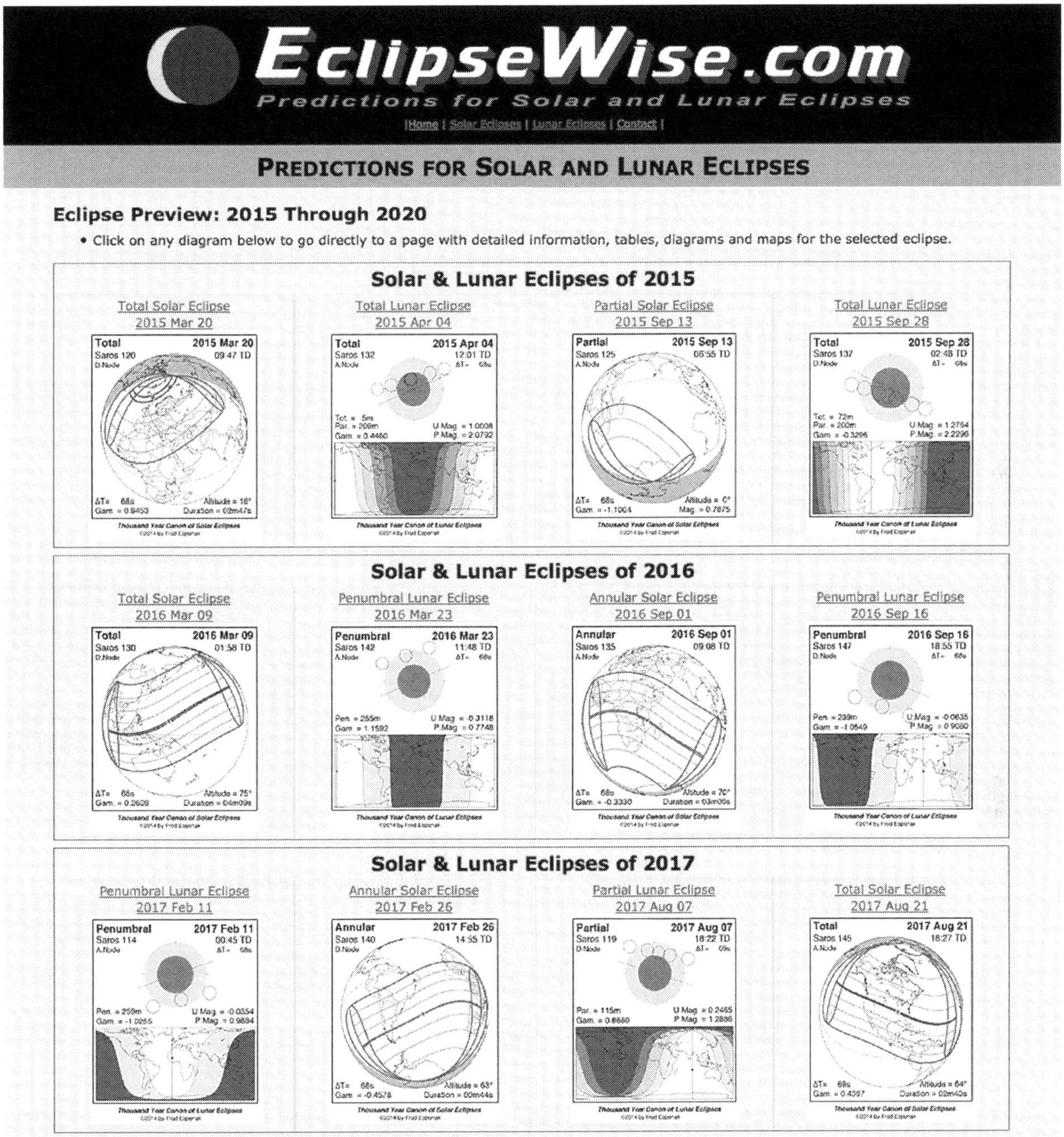

The Home Page of EclipseWise.com features a graphical interface to quickly preview upcoming eclipses with links to pages containing more detailed information.

The graphical user interface used by *EclipseWise.com* offers an intuitive way of accessing eclipse predictions. For example, the home page presents a concise preview of all upcoming solar and lunar eclipses over several years. Each small eclipse diagram gives a quick preview of an eclipse and links to a dedicated page for that particular eclipse.

The main or top pages of EclipseWise.com are:

Home Page (both solar and lunar eclipses): *eclipsewise.com/eclipse.html*
Solar Eclipses Page: *eclipsewise.com/solar/solar.html*
Lunar Eclipses Page: *eclipsewise.com/lunar/lunar.html*

EclipseWise.com and the 2017 Eclipse

EclipseWise.com has a series of pages and resources devoted to the 2017 eclipse. The main page is located at:

eclipsewise.com/solar/SEnews/TSE2017/TSE2017.html

It provides links to detailed eclipse path maps, tables of local eclipse circumstances for hundreds of cities, weather prospects along the eclipse path, and more. The link to an interactive Google Map with the eclipse path plotted on it allows the user to zoom into an part of the path. Click on any point on the map to display the eclipse circumstances and duration of totality at that location.

Other features include information on eye safety, eclipse photography, the sky during totality and additional data tables about the eclipse path. This web site will continue to add features as the eclipse approaches.

1.5 The 2017 Eclipse Bulletin

This atlas is a companion to the publication *Eclipse Bulletin: Total Solar Eclipse of 2017 August 21* (Espenak and Anderson). The *2017 Eclipse Bulletin* is the most comprehensive guide to the 2017 eclipse. It contains complete details and coordinates for the path of the Moon's shadow. Tables of local circumstances for more than 1000 cities across the USA provide times of each phase of the eclipse along with the eclipse magnitude, duration and the Sun's altitude. Additional tables cover cities in Canada, Mexico, Central and South America and Europe. An exhaustive climatological study identifies areas along the eclipse path where the highest probability of favorable weather may be found. A travelogue highlights key locations in the eclipse track from Oregon through South Carolina. Finally, detailed information is presented about solar filters and how to safely observe and photograph the eclipse.

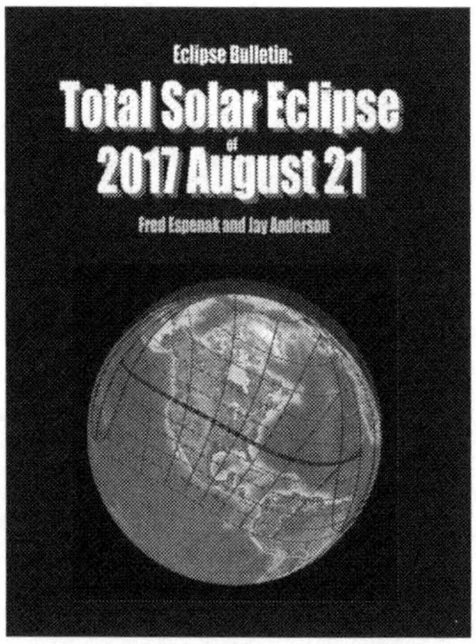

For more information, see: *astropixels.com/pubs/TSE2017.html*

1.6 Eclipse Predictions

The algorithms and software for the eclipse predictions were developed primarily from the *Explanatory Supplement to the Astronomical Ephemeris* (Her Majesty's Nautical Almanac Office, 1974) with additional algorithms from *Elements of Solar Eclipses: 1951–2200* (Meeus, 1989). The solar and lunar ephemerides were generated from the JPL DE405. All eclipse calculations were made using a value for the Moon's radius of k=0.2722810 for the path of totality. Center of mass coordinates for the Moon have been used without correction to the lunar limb profile. A value for ΔT of 68.6 seconds was used to convert the predictions from Terrestrial Dynamical Time to Universal Time (UT1).

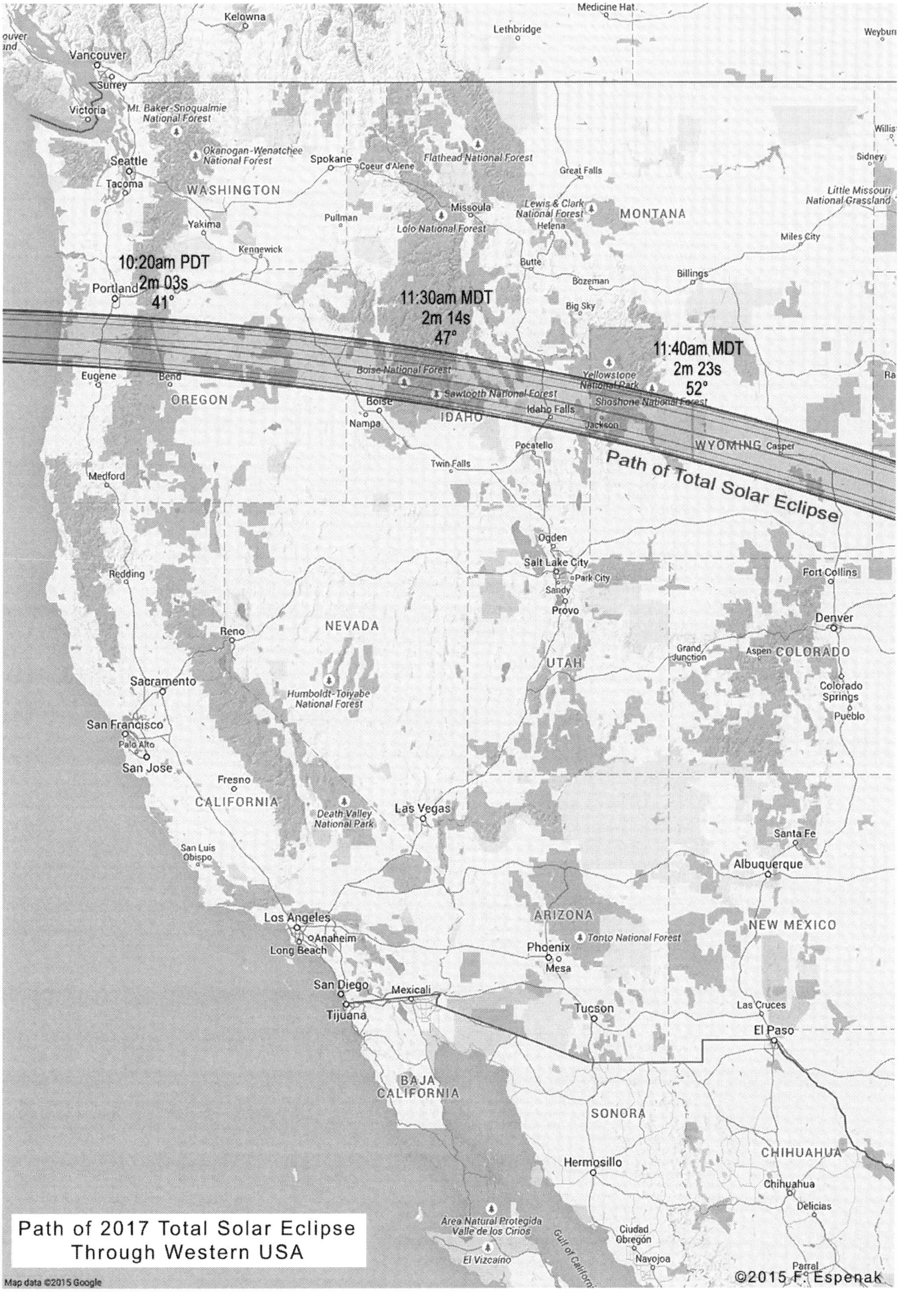

Path of 2017 Total Solar Eclipse
Through Western USA

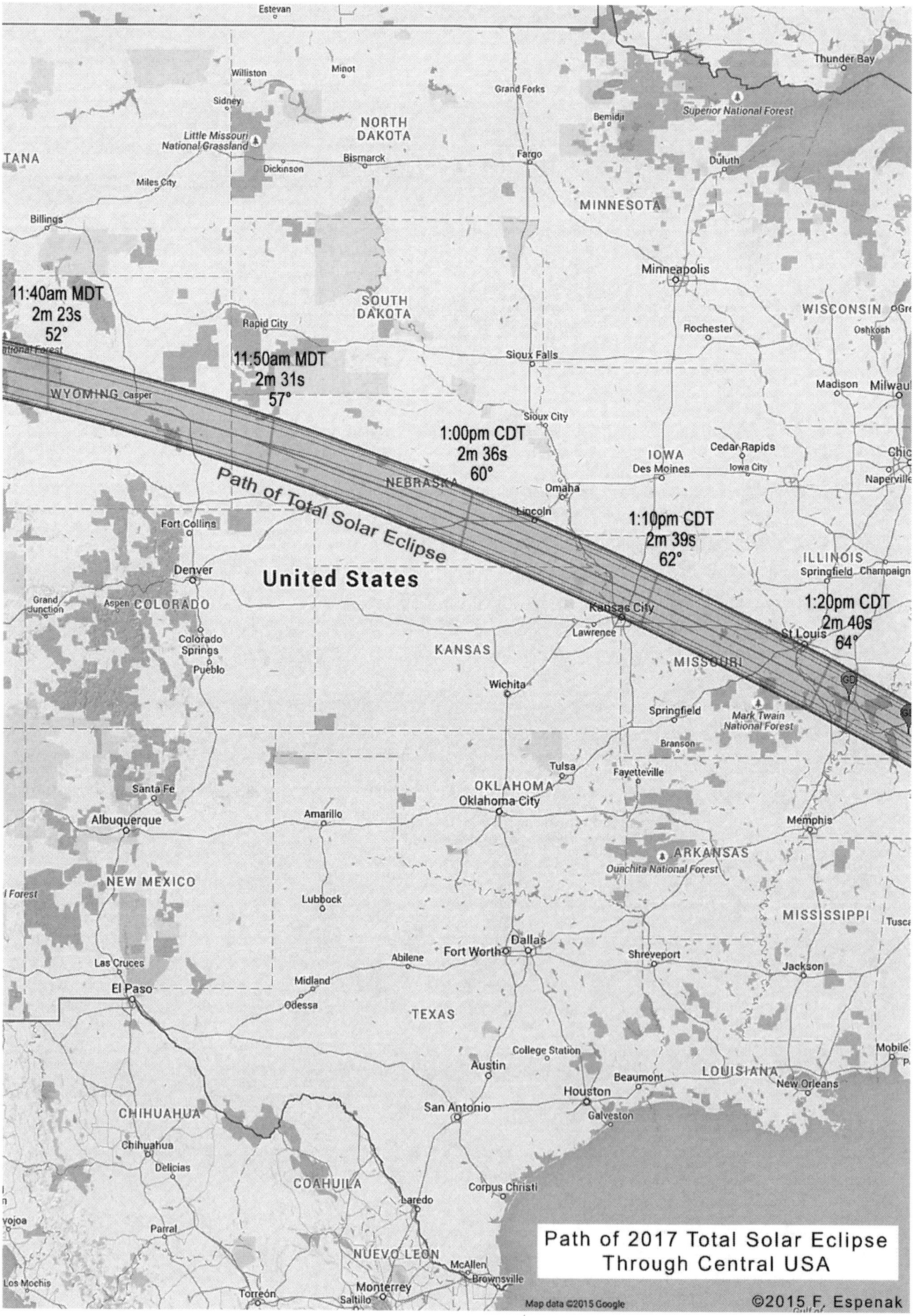
11:40am MDT
2m 23s
52°
11:50am MDT
2m 31s
57°
1:00pm CDT
2m 36s
60°
1:10pm CDT
2m 39s
62°
1:20pm CDT
2m 40s
64°
Path of Total Solar Eclipse
United States
NORTH DAKOTA
SOUTH DAKOTA
MINNESOTA
WISCONSIN
WYOMING
NEBRASKA
IOWA
ILLINOIS
COLORADO
KANSAS
MISSOURI
OKLAHOMA
ARKANSAS
NEW MEXICO
MISSISSIPPI
TEXAS
LOUISIANA
CHIHUAHUA
COAHUILA
NUEVO LEON
Path of 2017 Total Solar Eclipse
Through Central USA
Map data ©2015 Google
©2015 F. Espenak

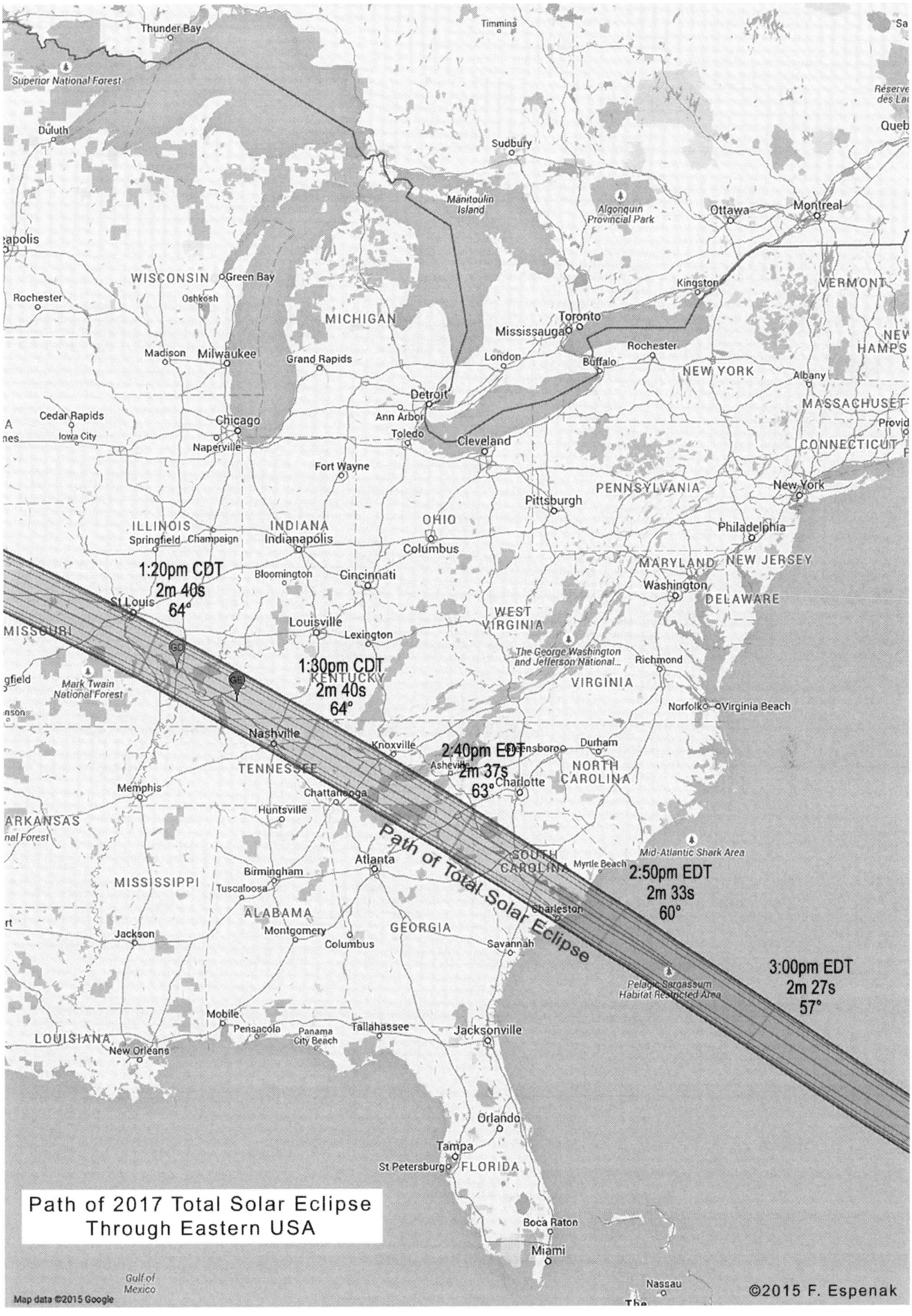
1:20pm CDT
2m 40s
64°
1:30pm CDT
2m 40s
64°
2:40pm EDT
2m 37s
63°
2:50pm EDT
2m 33s
60°
3:00pm EDT
2m 27s
57°
Path of Total Solar Eclipse
Path of 2017 Total Solar Eclipse
Through Eastern USA
©2015 F. Espenak
Map data ©2015 Google

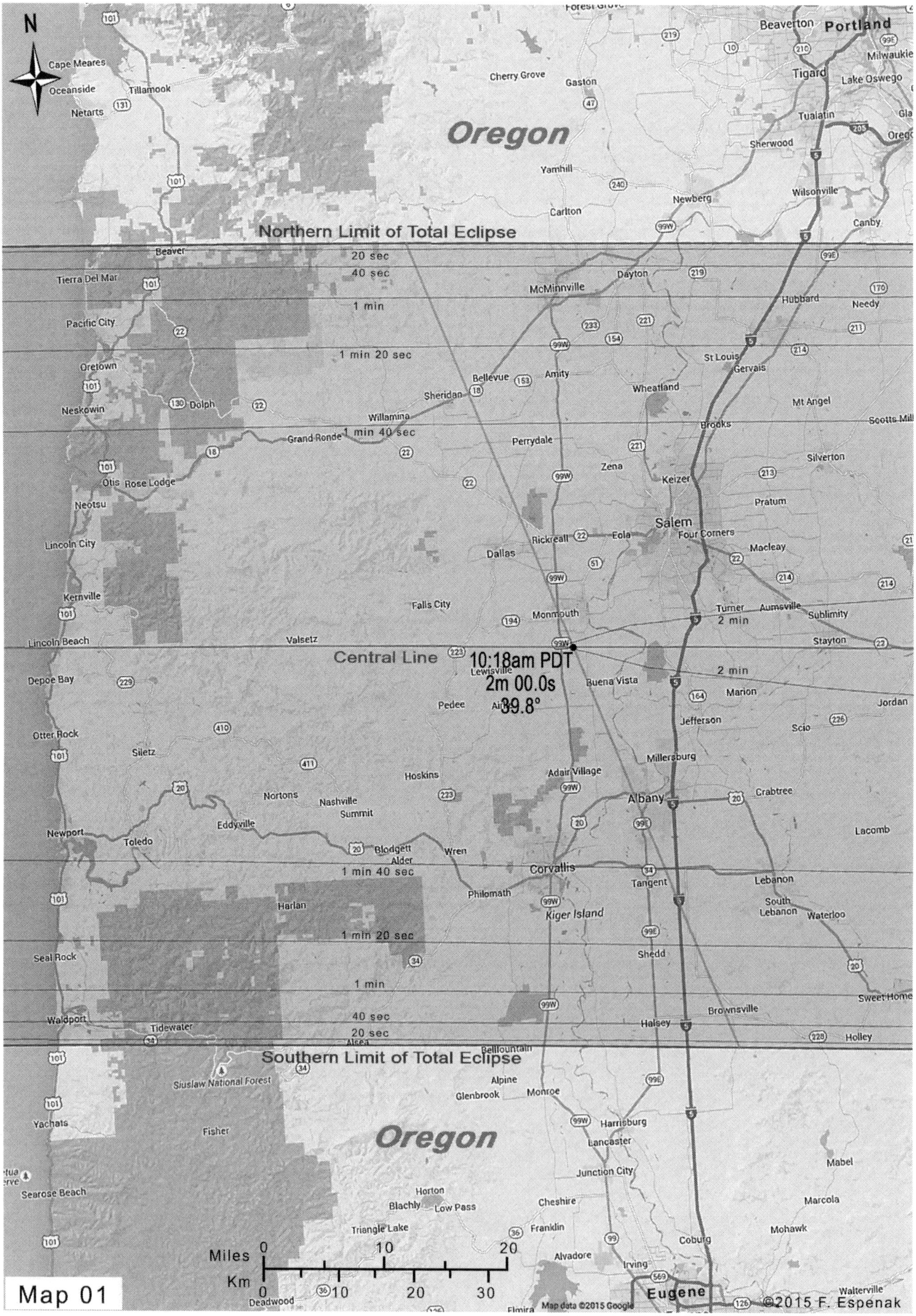
N
Oregon
Northern Limit of Total Eclipse
20 sec
40 sec
1 min
1 min 20 sec
1 min 40 sec
2 min
Central Line
10:18am PDT
2m 00.0s
39.8°
2 min
1 min 40 sec
1 min 20 sec
1 min
40 sec
20 sec
Southern Limit of Total Eclipse
Oregon
Portland
Beaverton
Tigard
Salem
Albany
Corvallis
Eugene
McMinnville
Newport
Lincoln City
Tillamook
Miles 0 10 20
Km 0 10 20 30
Map 01
Map data ©2015 Google
©2015 F. Espenak

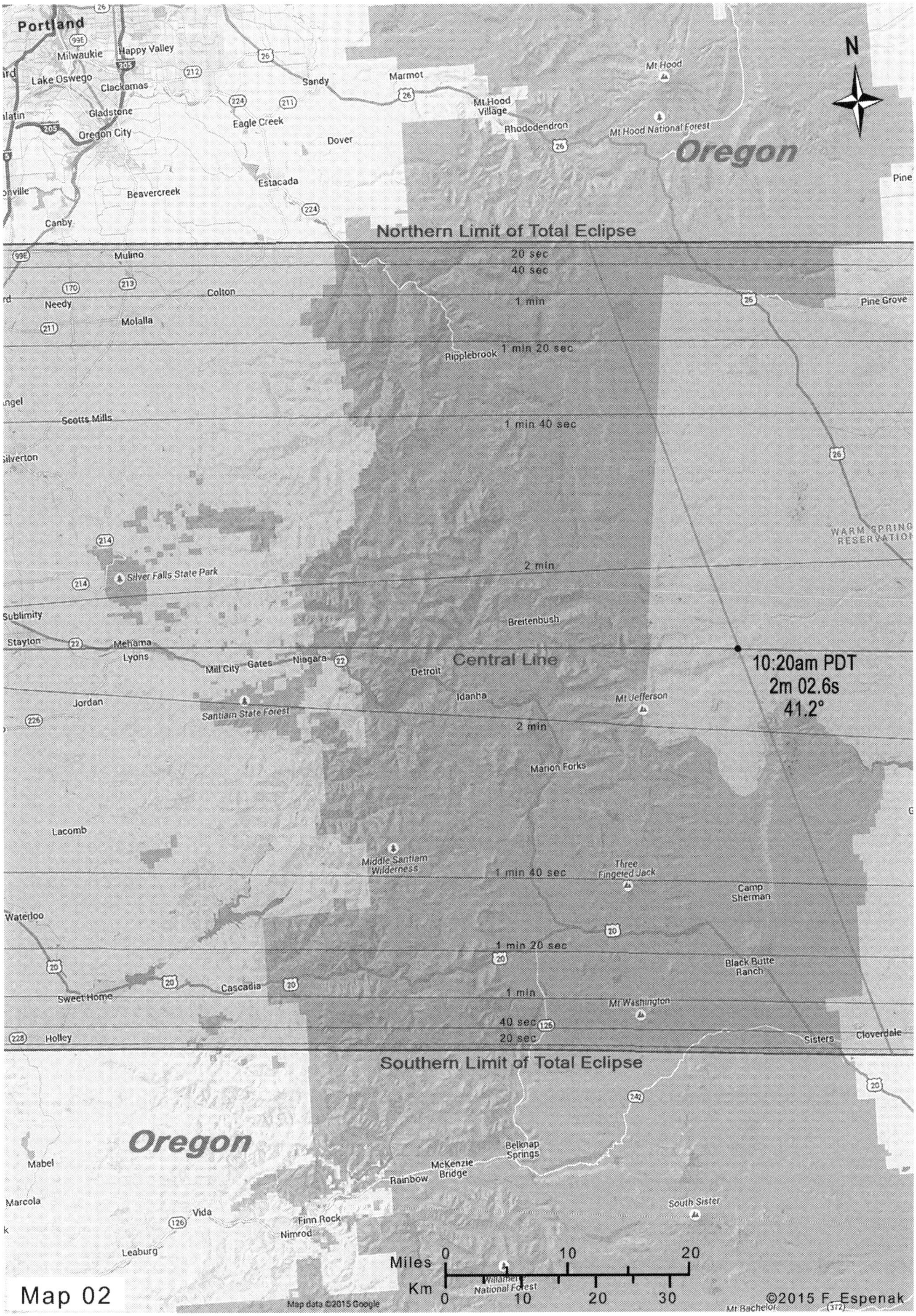
Northern Limit of Total Eclipse
20 sec
40 sec
1 min
1 min 20 sec
1 min 40 sec
2 min
Central Line
10:20am PDT
2m 02.6s
41.2°
2 min
1 min 40 sec
1 min 20 sec
1 min
40 sec
20 sec
Southern Limit of Total Eclipse
Oregon
Portland
Milwaukie
Happy Valley
Lake Oswego
Clackamas
Gladstone
Oregon City
Sandy
Marmot
Eagle Creek
Dover
Estacada
Beavercreek
Canby
Mt Hood
Mt Hood Village
Rhododendron
Mt Hood National Forest
Mulino
Colton
Needy
Molalla
Ripplebrook
Pine Grove
Scotts Mills
Silverton
Silver Falls State Park
Sublimity
Stayton
Mehama
Lyons
Mill City
Gates
Niagara
Detroit
Breitenbush
Idanha
Mt Jefferson
Jordan
Santiam State Forest
Marion Forks
Lacomb
Middle Santiam Wilderness
Three Fingered Jack
Camp Sherman
Waterloo
Sweet Home
Cascadia
Black Butte Ranch
Mt Washington
Holley
Sisters
Cloverdale
Belknap Springs
McKenzie Bridge
Rainbow
Mabel
Marcola
Vida
Finn Rock
Nimrod
Leaburg
South Sister
Willamette National Forest
Mt Bachelor
WARM SPRINGS RESERVATION
N
Miles 0 10 20
Km 0 10 20 30
Map 02
Map data ©2015 Google
©2015 F. Espenak

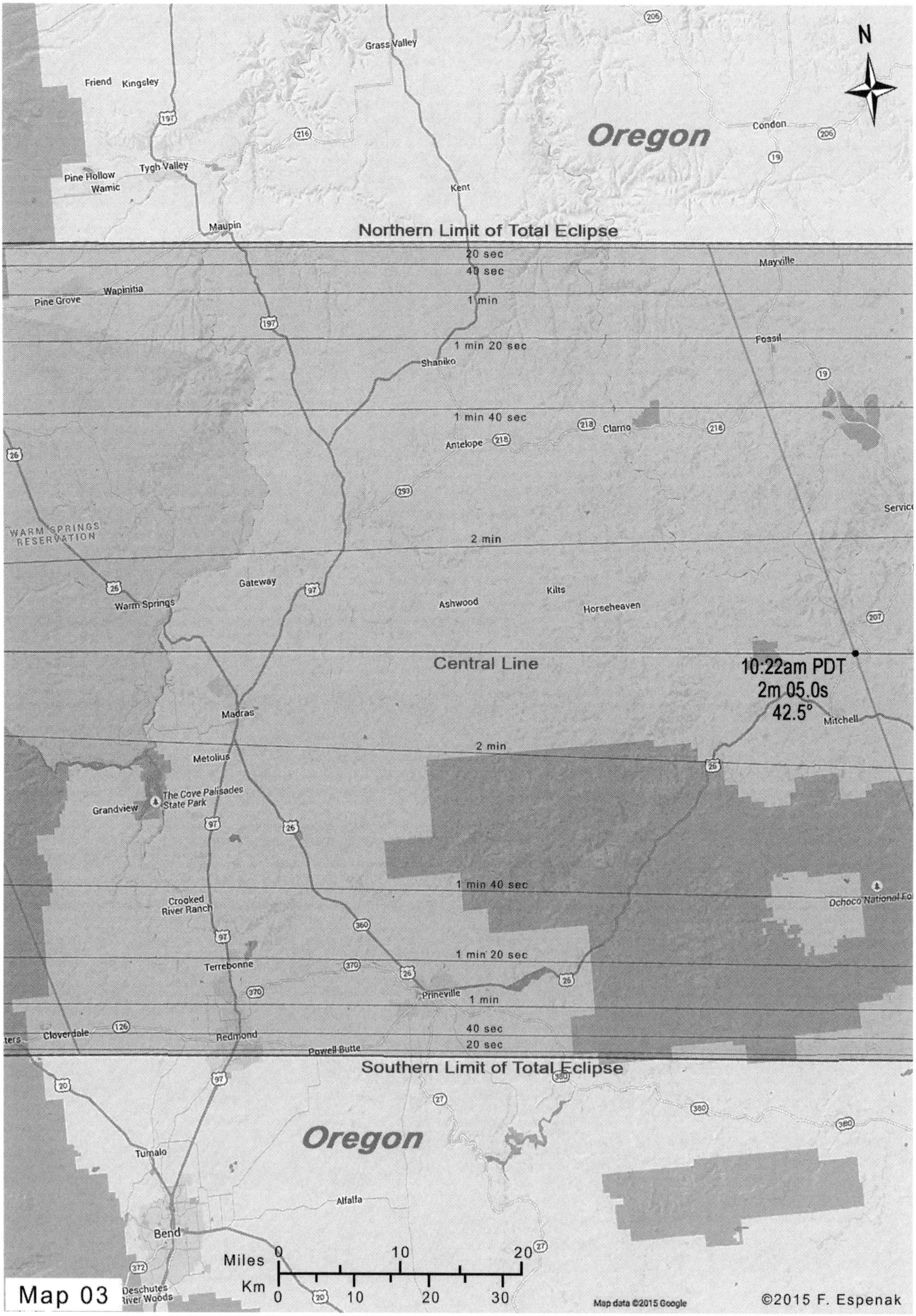
N
Oregon
Grass Valley
Friend
Kingsley
Condon
Pine Hollow
Wamic
Tygh Valley
Kent
Maupin
Northern Limit of Total Eclipse
20 sec
40 sec
Mayville
Pine Grove
Wapinitia
1 min
Fossil
1 min 20 sec
Shaniko
1 min 40 sec
Clarno
Antelope
Service
WARM SPRINGS RESERVATION
2 min
Gateway
Warm Springs
Ashwood
Kilts
Horseheaven
Central Line
10:22am PDT
2m 05.0s
42.5°
Madras
Mitchell
2 min
Metolius
The Cove Palisades State Park
Grandview
1 min 40 sec
Ochoco National Fo
Crooked River Ranch
1 min 20 sec
Terrebonne
Prineville
1 min
Cloverdale
Redmond
40 sec
20 sec
Powell Butte
Southern Limit of Total Eclipse
Oregon
Tumalo
Alfalfa
Bend
Miles
0
10
20
Km
0
10
20
30
Map 03
Deschutes River Woods
Map data ©2015 Google
©2015 F. Espenak

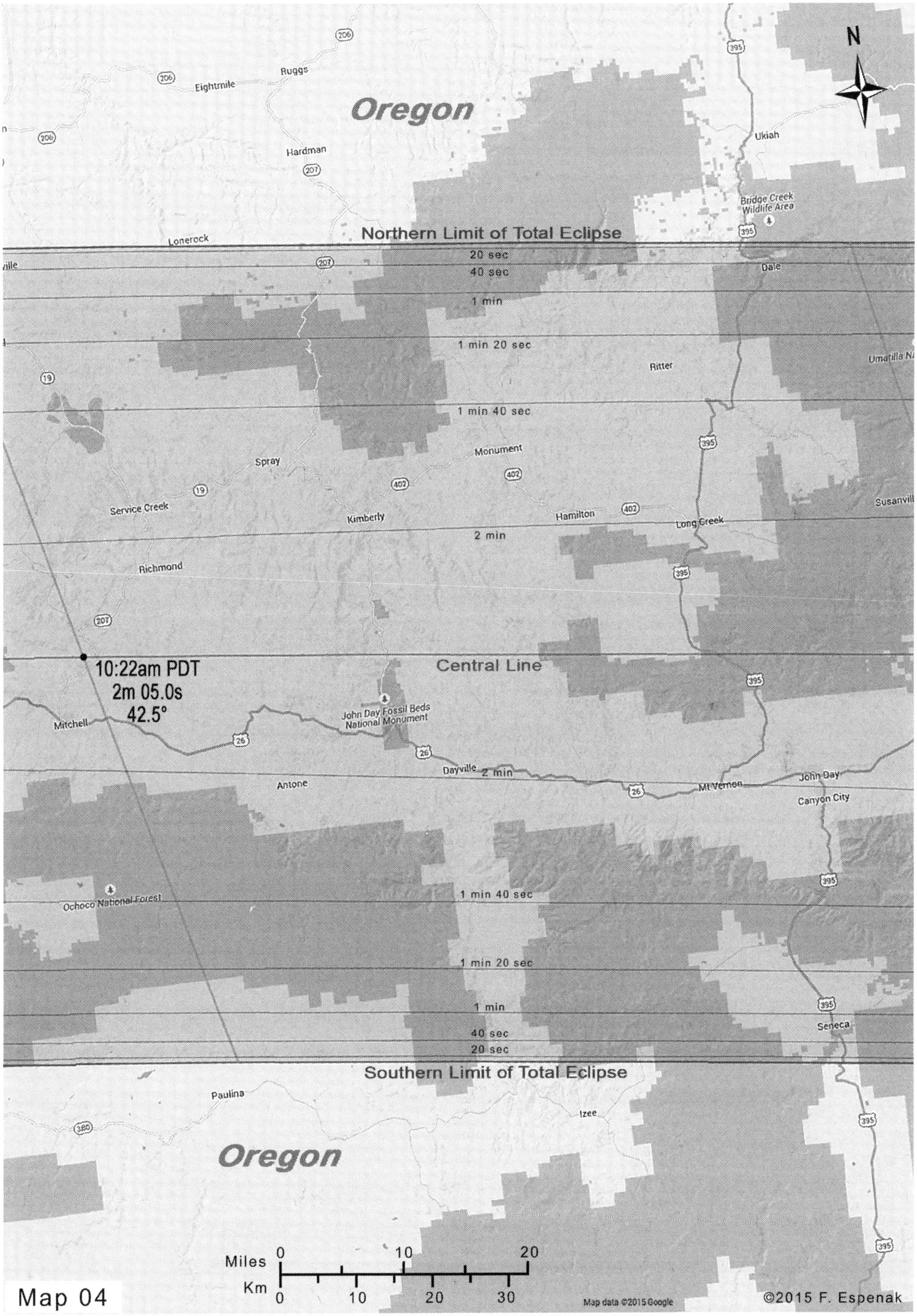

Oregon
Ruggs
Eightmile
Hardman
Ukiah
Bridge Creek Wildlife Area
Northern Limit of Total Eclipse
Lonerock
20 sec
40 sec
Dale
1 min
1 min 20 sec
Ritter
Umatilla N
1 min 40 sec
Monument
Spray
Service Creek
Kimberly
Hamilton
Long Creek
Susanvill
2 min
Richmond
10:22am PDT
2m 05.0s
42.5°
Central Line
John Day Fossil Beds National Monument
Mitchell
Dayville
2 min
Antone
Mt Vernon
John Day
Canyon City
Ochoco National Forest
1 min 40 sec
1 min 20 sec
1 min
Seneca
40 sec
20 sec
Southern Limit of Total Eclipse
Paulina
Izee
Oregon
Miles
Km
0
10
20
30
Map 04
Map data ©2015 Google
©2015 F. Espenak

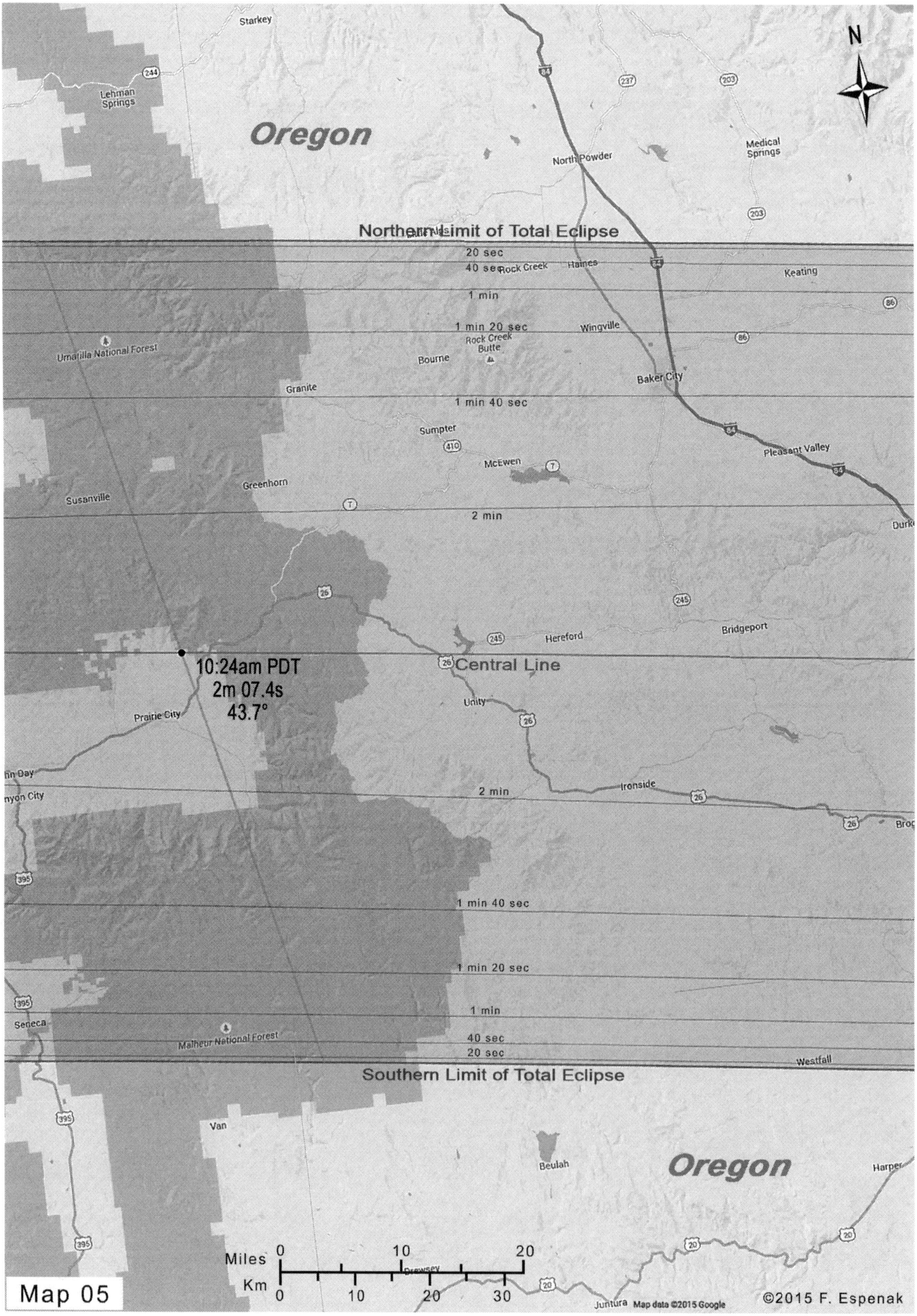
N
Oregon
Starkey
Lehman Springs
North Powder
Medical Springs
Northern Limit of Total Eclipse
20 sec
40 sec
Rock Creek
Haines
Keating
1 min
1 min 20 sec
Wingville
Rock Creek Butte
Bourne
Umatilla National Forest
Baker City
Granite
1 min 40 sec
Sumpter
McEwen
Pleasant Valley
Greenhorn
Susanville
2 min
Hereford
Bridgeport
10:24am PDT
2m 07.4s
43.7°
Central Line
Unity
Prairie City
Ironside
2 min
1 min 40 sec
1 min 20 sec
1 min
Seneca
Malheur National Forest
40 sec
20 sec
Westfall
Southern Limit of Total Eclipse
Van
Beulah
Oregon
Harper
Miles
0
10
20
Km
0
10
20
30
Drewsey
Juntura
Map data ©2015 Google
Map 05
©2015 F. Espenak

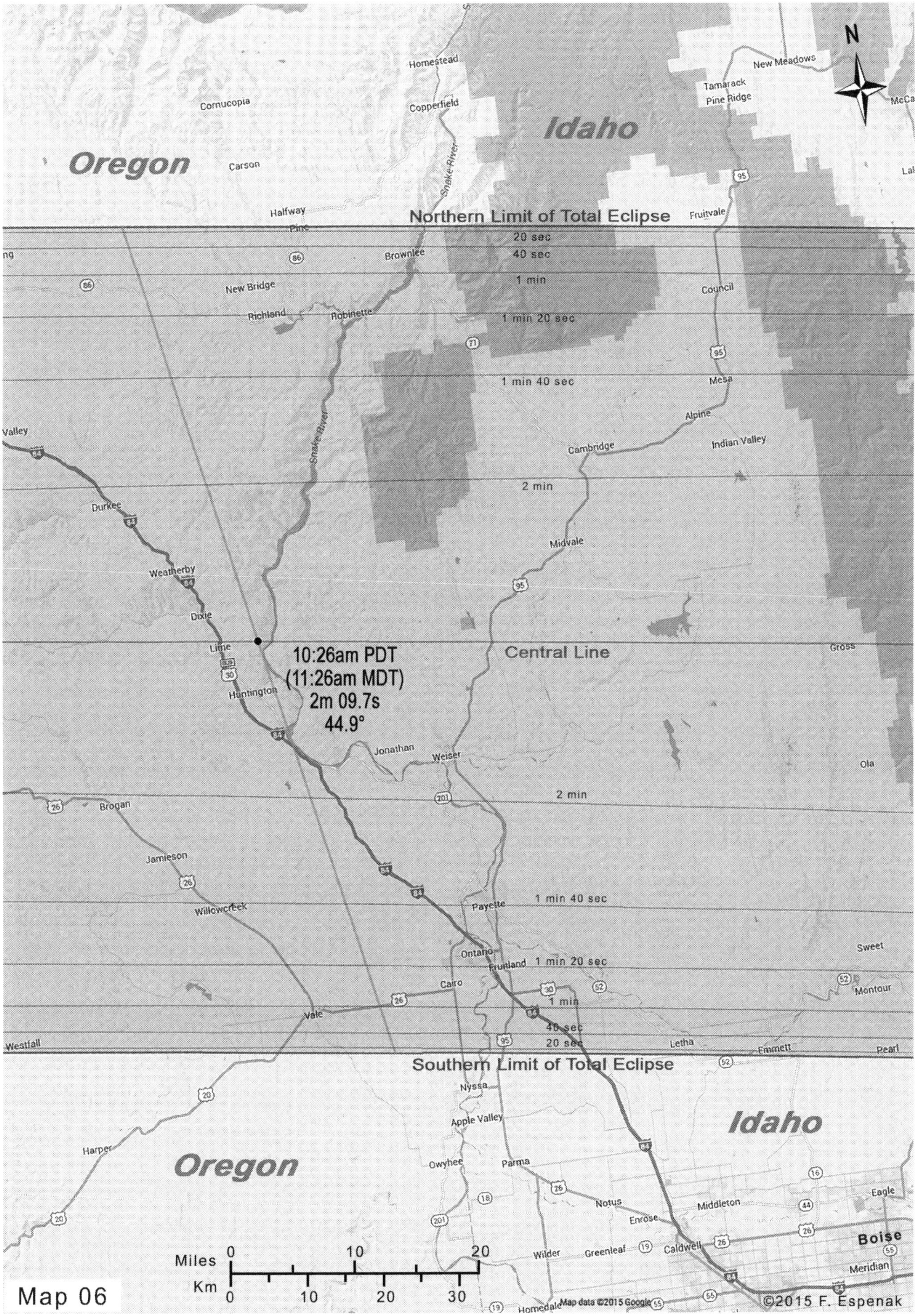
N
Idaho
Oregon
Homestead
Cornucopia
Copperfield
New Meadows
Tamarack
Pine Ridge
Carson
Halfway
Pine
Northern Limit of Total Eclipse
Fruitvale
20 sec
40 sec
1 min
1 min 20 sec
1 min 40 sec
Brownlee
New Bridge
Richland
Robinette
Council
Mesa
Alpine
Indian Valley
Cambridge
Valley
Snake River
Durkee
2 min
Midvale
Weatherby
Dixie
Lime
Central Line
Gross
10:26am PDT
(11:26am MDT)
2m 09.7s
44.9°
Huntington
Jonathan
Weiser
Ola
Brogan
2 min
Jamieson
Willowcreek
1 min 40 sec
Payette
Ontario
Fruitland
1 min 20 sec
Sweet
Cairo
1 min
Montour
Vale
40 sec
20 sec
Westfall
Letha
Emmett
Pearl
Southern Limit of Total Eclipse
Nyssa
Apple Valley
Idaho
Harper
Oregon
Owyhee
Parma
Notus
Middleton
Eagle
Enrose
Boise
Miles
0
10
20
Km
0
10
20
30
Wilder
Greenleaf
Caldwell
Meridian
Map 06
Homedale
Map data ©2015 Google
©2015 F. Espenak

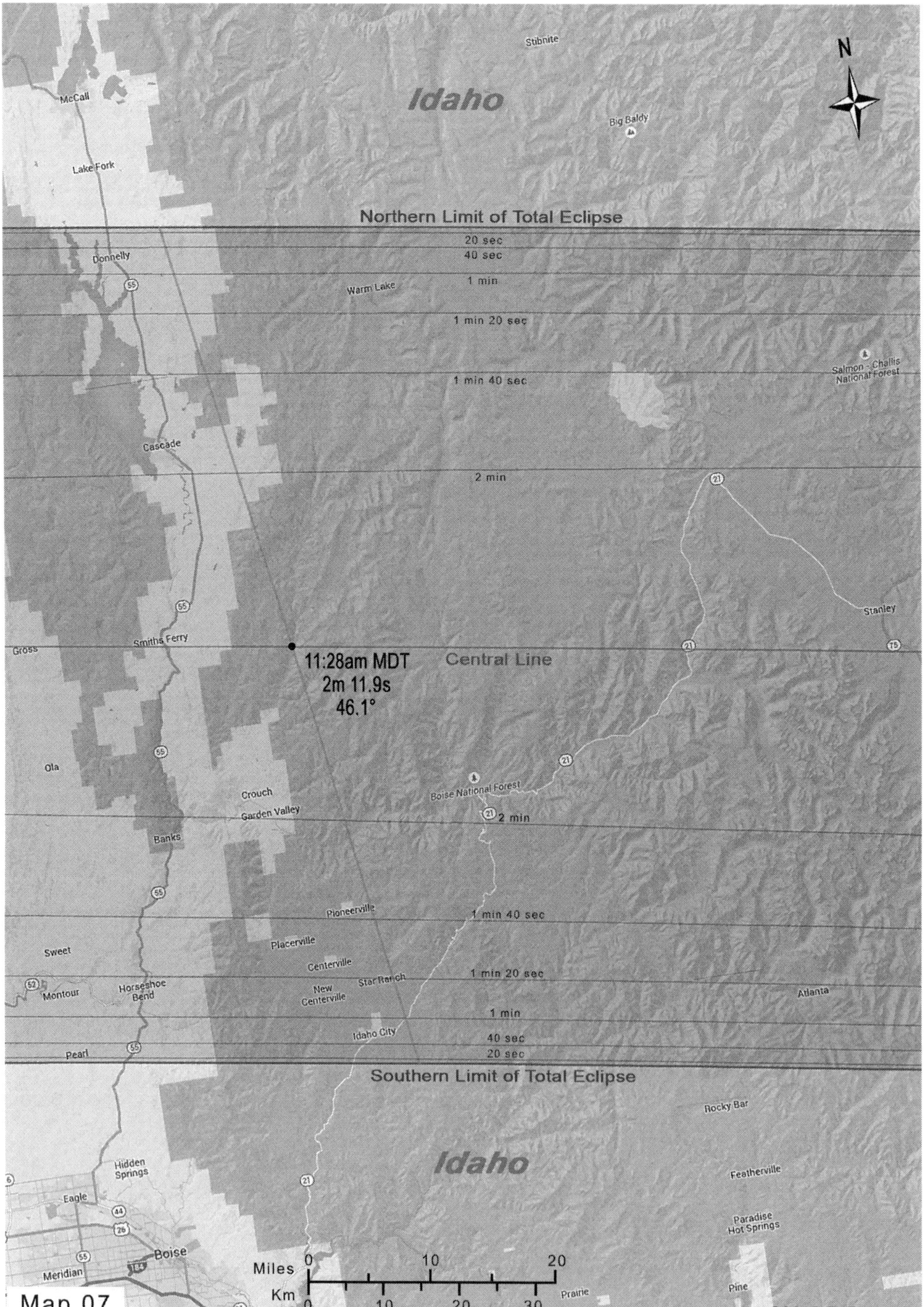
Idaho
Stibnite
N
McCall
Big Baldy
Lake Fork
Northern Limit of Total Eclipse
20 sec
40 sec
Donnelly
Warm Lake
1 min
1 min 20 sec
Salmon - Challis National Forest
1 min 40 sec
Cascade
2 min
Stanley
Gross
Smiths Ferry
11:28am MDT
2m 11.9s
46.1°
Central Line
Ola
Crouch
Garden Valley
Boise National Forest
2 min
Banks
Pioneerville
1 min 40 sec
Sweet
Placerville
Centerville
Montour
Horseshoe Bend
New Centerville
Star Ranch
1 min 20 sec
Atlanta
1 min
Idaho City
40 sec
Pearl
20 sec
Southern Limit of Total Eclipse
Rocky Bar
Idaho
Hidden Springs
Featherville
Eagle
Paradise Hot Springs
Boise
Meridian
Miles
0
10
20
Km
0
10
20
30
Prairie
Pine
Map 07
Map data ©2015 Google
©2015 F. Espenak

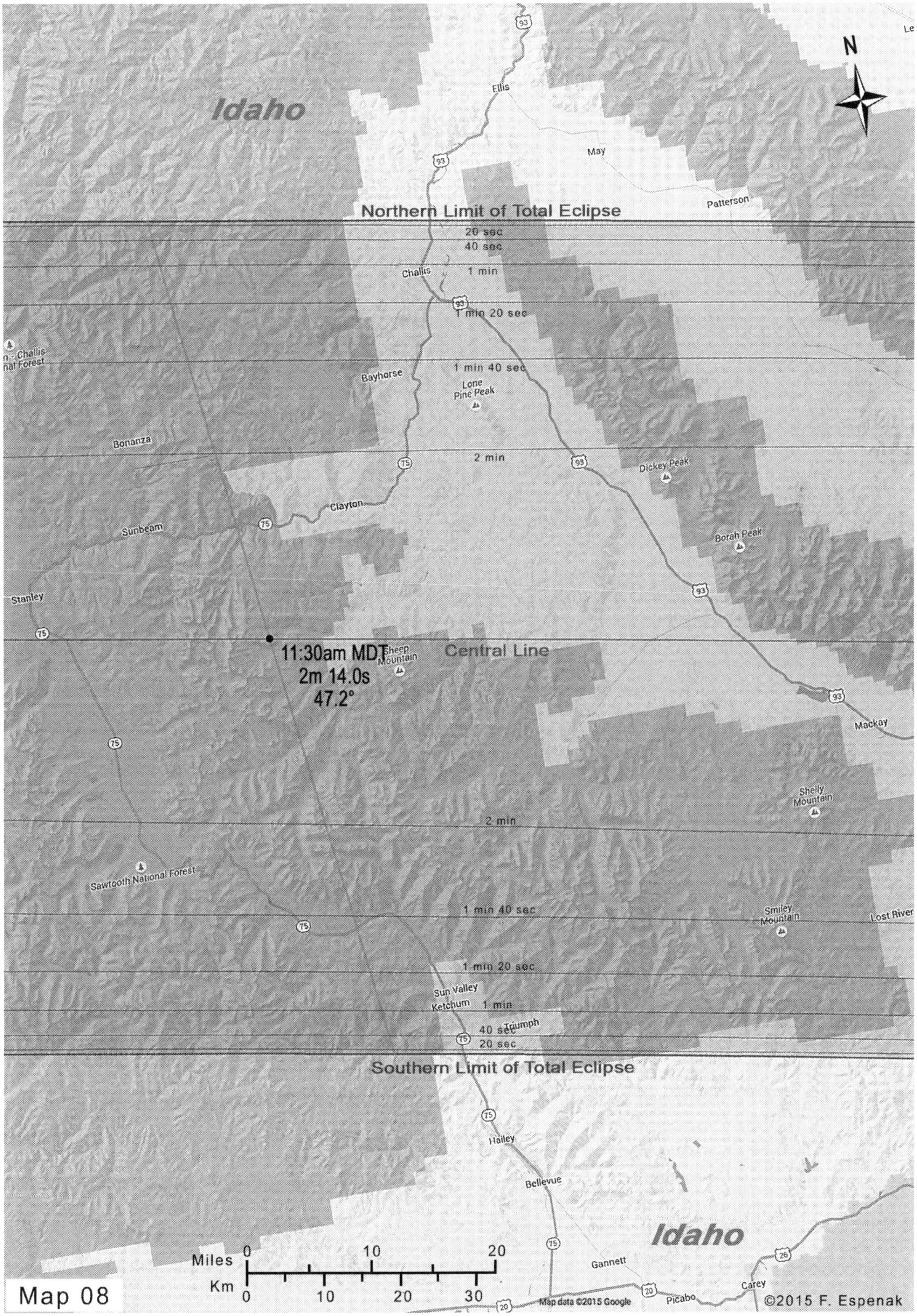

Idaho
N
Ellis
May
Patterson
Northern Limit of Total Eclipse
20 sec
40 sec
Challis
1 min
1 min 20 sec
1 min 40 sec
Bayhorse
Lone Pine Peak
Bonanza
2 min
Dickey Peak
Clayton
Sunbeam
Borah Peak
Stanley
11:30am MDT
2m 14.0s
47.2°
Sheep Mountain
Central Line
Mackay
Shelly Mountain
2 min
Sawtooth National Forest
1 min 40 sec
Smiley Mountain
Lost River
1 min 20 sec
Sun Valley
Ketchum
1 min
40 sec
Triumph
20 sec
Southern Limit of Total Eclipse
Hailey
Bellevue
Idaho
Miles 0 10 20
Km 0 10 20 30
Gannett
Carey
Picabo
Map data ©2015 Google
Map 08
©2015 F. Espenak

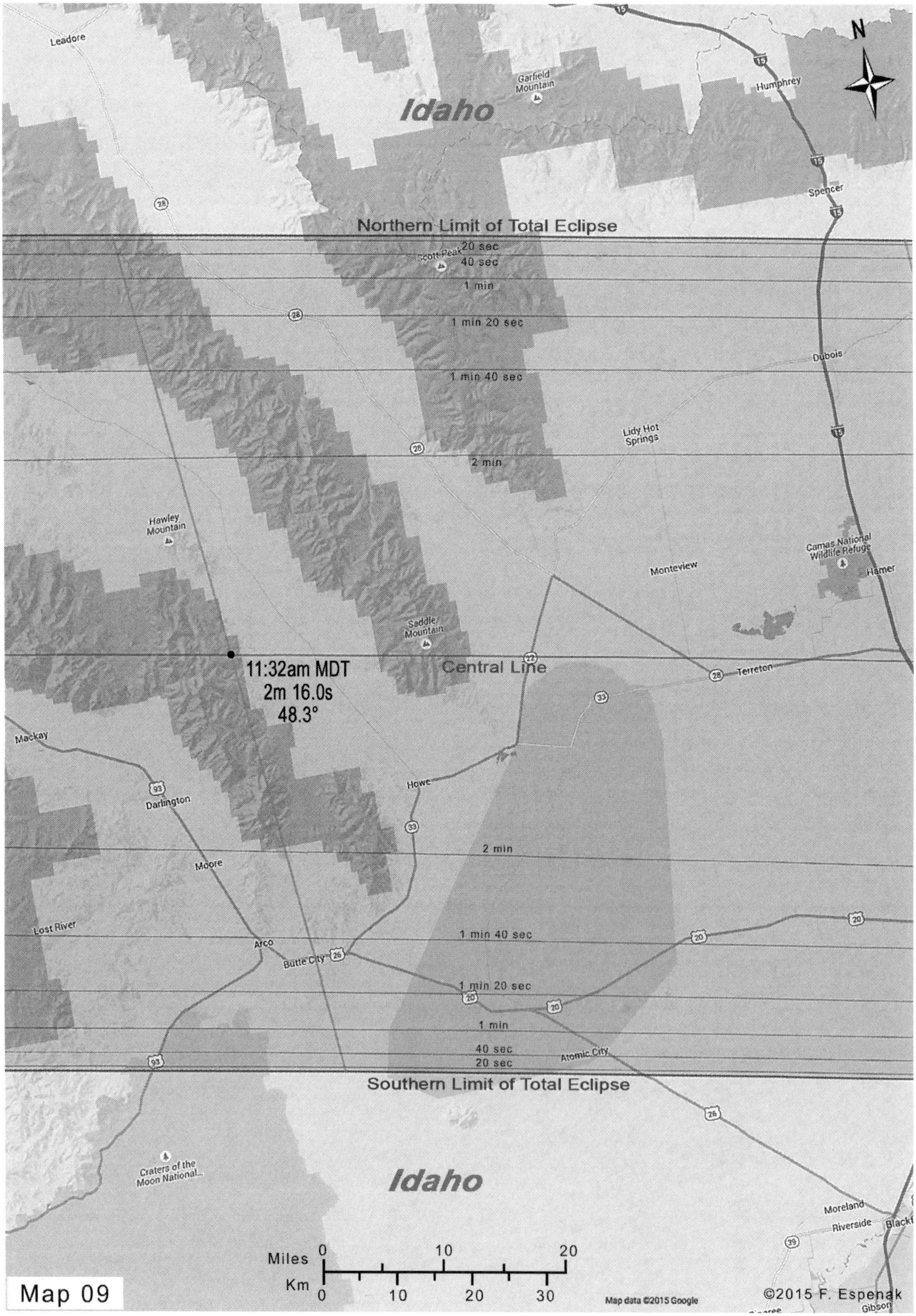
Leadore
Garfield Mountain
Idaho
Humphrey
N
Spencer
Northern Limit of Total Eclipse
Scott Peak
20 sec
40 sec
1 min
1 min 20 sec
Dubois
1 min 40 sec
Lidy Hot Springs
2 min
Hawley Mountain
Camas National Wildlife Refuge
Monteview
Hamer
Saddle Mountain
11:32am MDT
2m 16.0s
48.3°
Central Line
Terreton
Mackay
Howe
Darlington
2 min
Moore
Lost River
1 min 40 sec
Arco
Butte City
1 min 20 sec
1 min
40 sec
20 sec
Atomic City
Southern Limit of Total Eclipse
Craters of the Moon National...
Idaho
Moreland
Riverside
Miles 0 10 20
Km 0 10 20 30
Map 09
Map data ©2015 Google
©2015 F. Espenak
Gibson

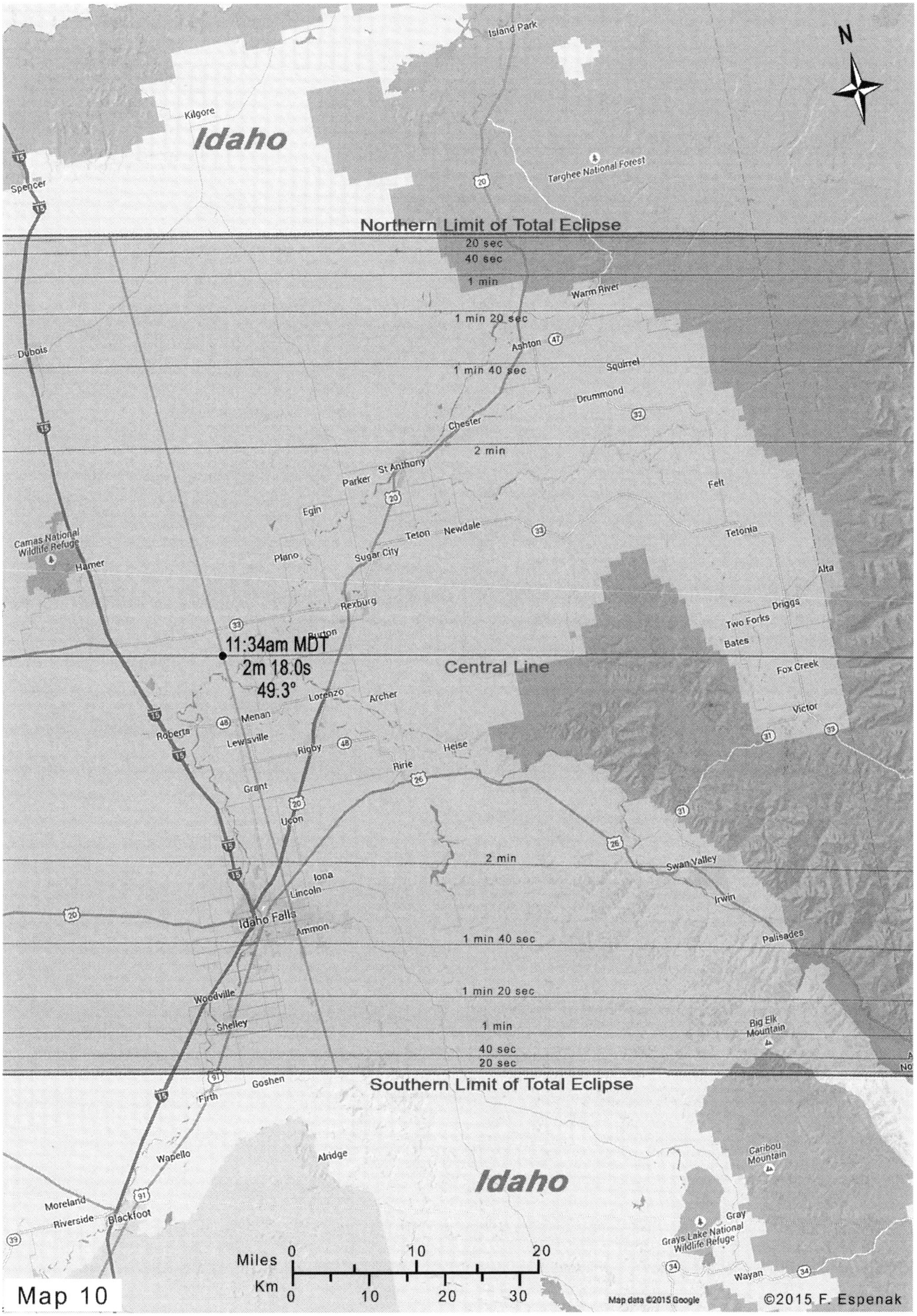
N
Idaho
Island Park
Kilgore
Spencer
Targhee National Forest
Northern Limit of Total Eclipse
20 sec
40 sec
1 min
Warm River
1 min 20 sec
Ashton
Dubois
1 min 40 sec
Squirrel
Drummond
Chester
2 min
St Anthony
Parker
Felt
Egin
Teton
Newdale
Tetonia
Camas National Wildlife Refuge
Hamer
Plano
Sugar City
Alta
Rexburg
Driggs
Two Forks
Burton
Bates
11:34am MDT
2m 18.0s
49.3°
Central Line
Fox Creek
Lorenzo
Archer
Menan
Victor
Roberts
Lewisville
Rigby
Heise
Ririe
Grant
Ucon
2 min
Swan Valley
Iona
Lincoln
Irwin
Idaho Falls
Ammon
Palisades
1 min 40 sec
1 min 20 sec
Woodville
1 min
Shelley
Big Elk Mountain
40 sec
20 sec
Goshen
Southern Limit of Total Eclipse
Firth
Wapello
Alridge
Caribou Mountain
Idaho
Moreland
Riverside
Blackfoot
Gray
Grays Lake National Wildlife Refuge
Miles 0 10 20
Km 0 10 20 30
Wayan
Map 10
Map data ©2015 Google
©2015 F. Espenak

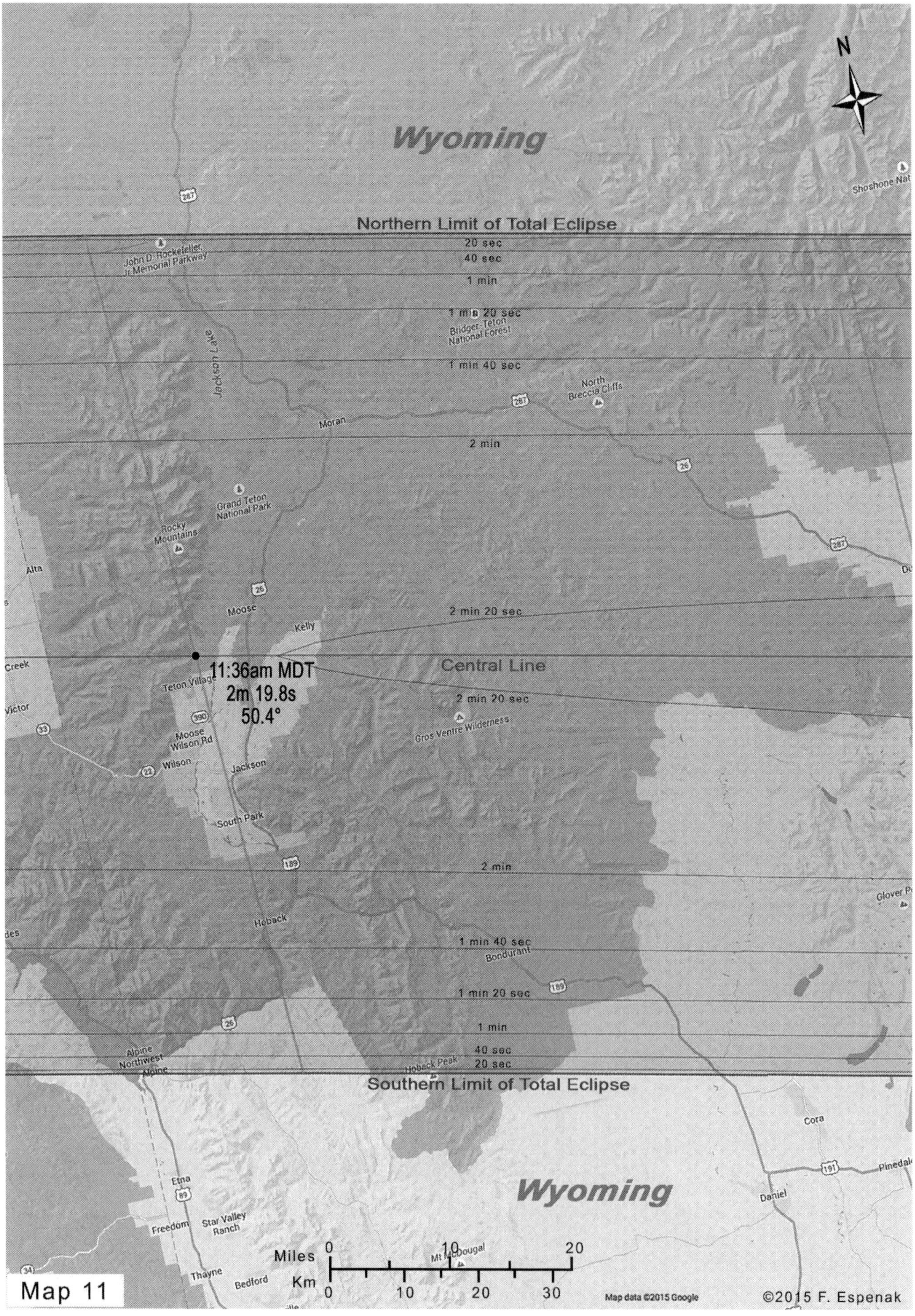
N
Wyoming
Shoshone Nat
287
Northern Limit of Total Eclipse
20 sec
40 sec
John D. Rockefeller, Jr. Memorial Parkway
1 min
1 min 20 sec
Bridger-Teton National Forest
Jackson Lake
1 min 40 sec
North Breccia Cliffs
287
Moran
2 min
26
Grand Teton National Park
Rocky Mountains
287
Alta
26
Moose
2 min 20 sec
Kelly
Creek
11:36am MDT
2m 19.8s
50.4°
Central Line
Teton Village
Victor
2 min 20 sec
390
Moose Wilson Rd
Gros Ventre Wilderness
33
22
Wilson
Jackson
South Park
189
2 min
Glover P
Hoback
1 min 40 sec
Bondurant
189
1 min 20 sec
26
1 min
Alpine Northwest
40 sec
Alpine
Hoback Peak
20 sec
Southern Limit of Total Eclipse
Cora
191
Pinedale
Etna
89
Wyoming
Daniel
Freedom
Star Valley Ranch
Miles 0 10 20
Mt McDougal
Km 0 10 20 30
34
Thayne
Bedford
Map 11
Map data ©2015 Google
©2015 F. Espenak

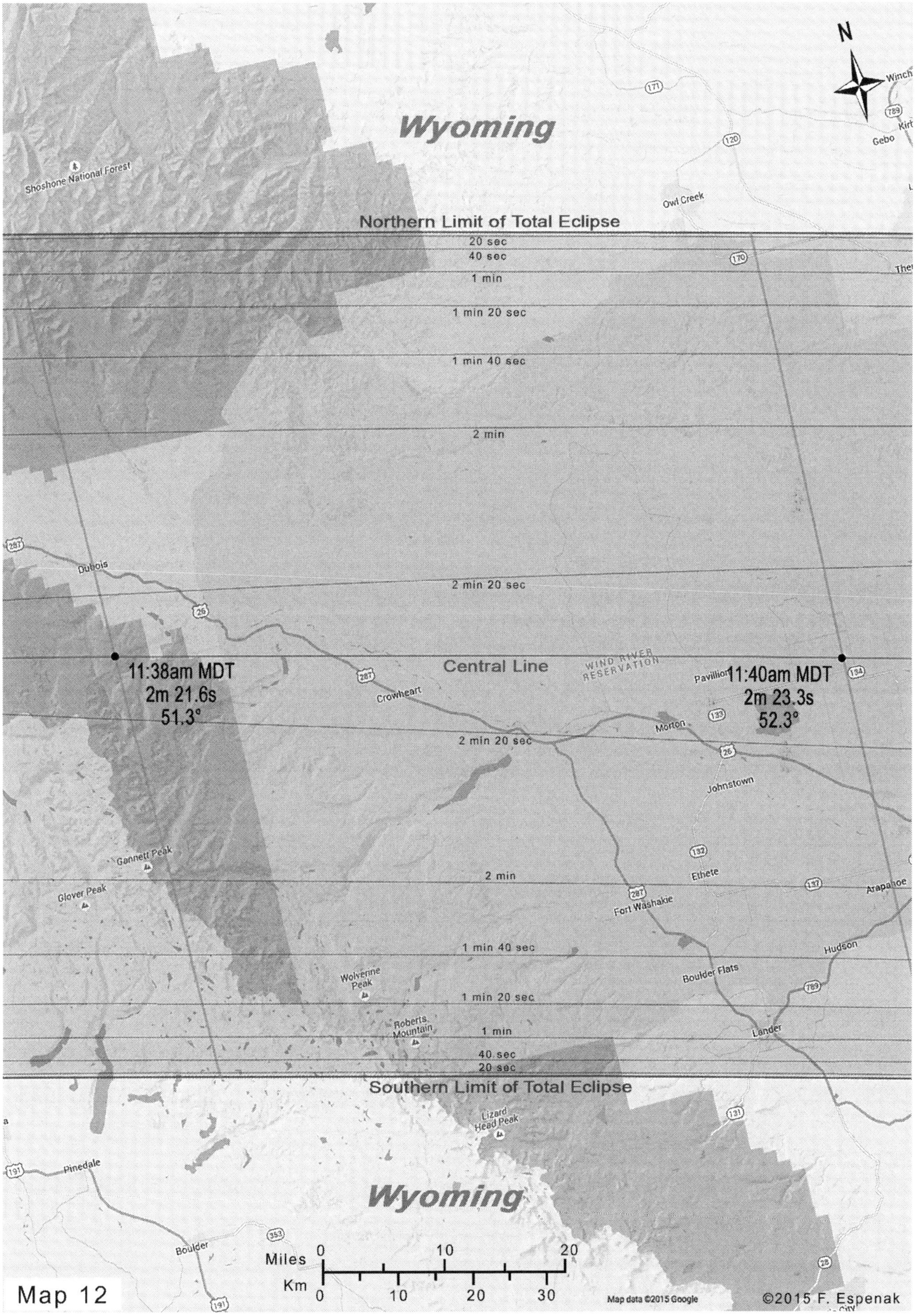

N
Wyoming
Shoshone National Forest
Owl Creek
Gebo
Northern Limit of Total Eclipse
20 sec
40 sec
1 min
1 min 20 sec
1 min 40 sec
2 min
2 min 20 sec
Dubois
Central Line
WIND RIVER RESERVATION
11:38am MDT
2m 21.6s
51.3°
Pavillion
11:40am MDT
2m 23.3s
52.3°
Crowheart
Morton
Johnstown
Ethete
Arapahoe
Fort Washakie
Gannett Peak
Glover Peak
Hudson
Boulder Flats
Wolverine Peak
Roberts Mountain
Lander
Southern Limit of Total Eclipse
Lizard Head Peak
Pinedale
Boulder
Miles 0 10 20
Km 0 10 20 30
Map 12
Map data ©2015 Google
©2015 F. Espenak

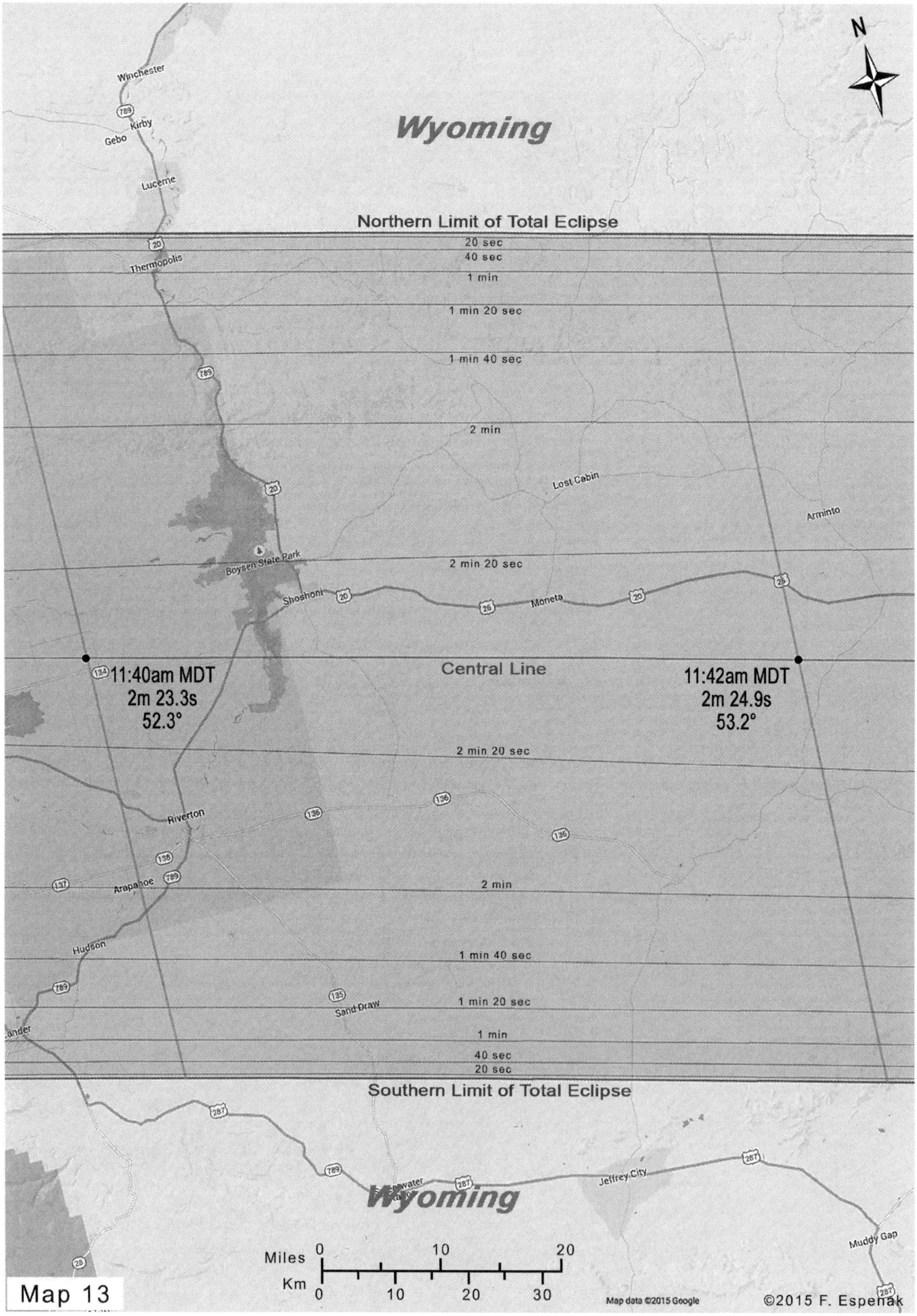
N
Wyoming
Winchester
Gebo
Kirby
Lucerne
Northern Limit of Total Eclipse
20 sec
40 sec
1 min
1 min 20 sec
1 min 40 sec
2 min
Thermopolis
Lost Cabin
Arminto
Boysen State Park
2 min 20 sec
Shoshoni
Moneta
Central Line
11:40am MDT
2m 23.3s
52.3°
11:42am MDT
2m 24.9s
53.2°
2 min 20 sec
Riverton
Arapahoe
2 min
Hudson
1 min 40 sec
Sand Draw
1 min 20 sec
Lander
1 min
40 sec
20 sec
Southern Limit of Total Eclipse
Jeffrey City
Wyoming
Muddy Gap
Miles 0 10 20
Km 0 10 20 30
Map data ©2015 Google
©2015 F. Espenak
Map 13

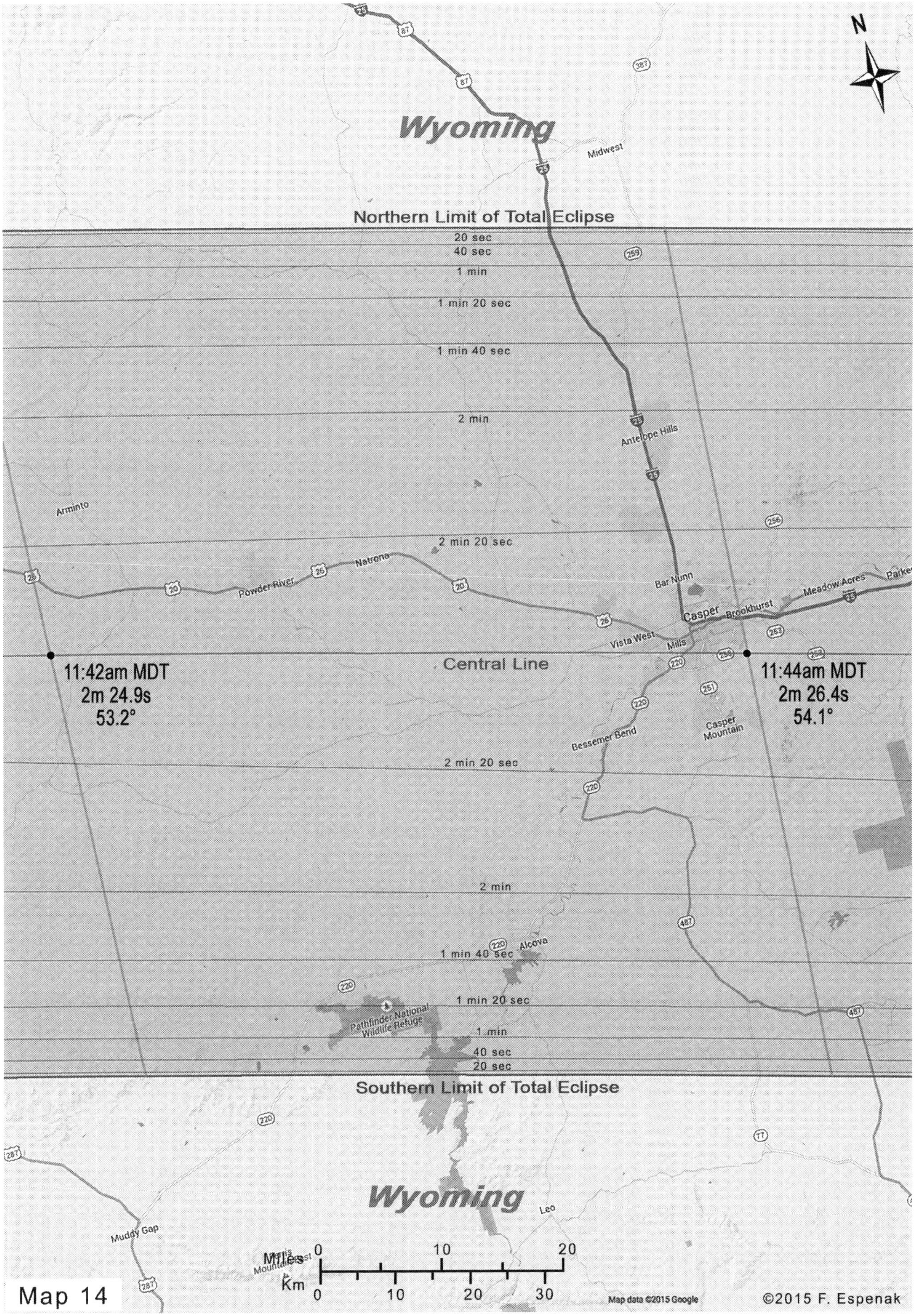

N
Wyoming
Midwest
Northern Limit of Total Eclipse
20 sec
40 sec
1 min
1 min 20 sec
1 min 40 sec
2 min
Antelope Hills
Arminto
2 min 20 sec
Natrona
Powder River
Bar Nunn
Meadow Acres
Parker
Casper
Brookhurst
Vista West
Mills
11:42am MDT
2m 24.9s
53.2°
Central Line
11:44am MDT
2m 26.4s
54.1°
Casper Mountain
Bessemer Bend
2 min 20 sec
2 min
Alcova
1 min 40 sec
1 min 20 sec
Pathfinder National Wildlife Refuge
1 min
40 sec
20 sec
Southern Limit of Total Eclipse
Wyoming
Leo
Muddy Gap
Miles 0 10 20
Km 0 10 20 30
Map data ©2015 Google
©2015 F. Espenak

Map 14

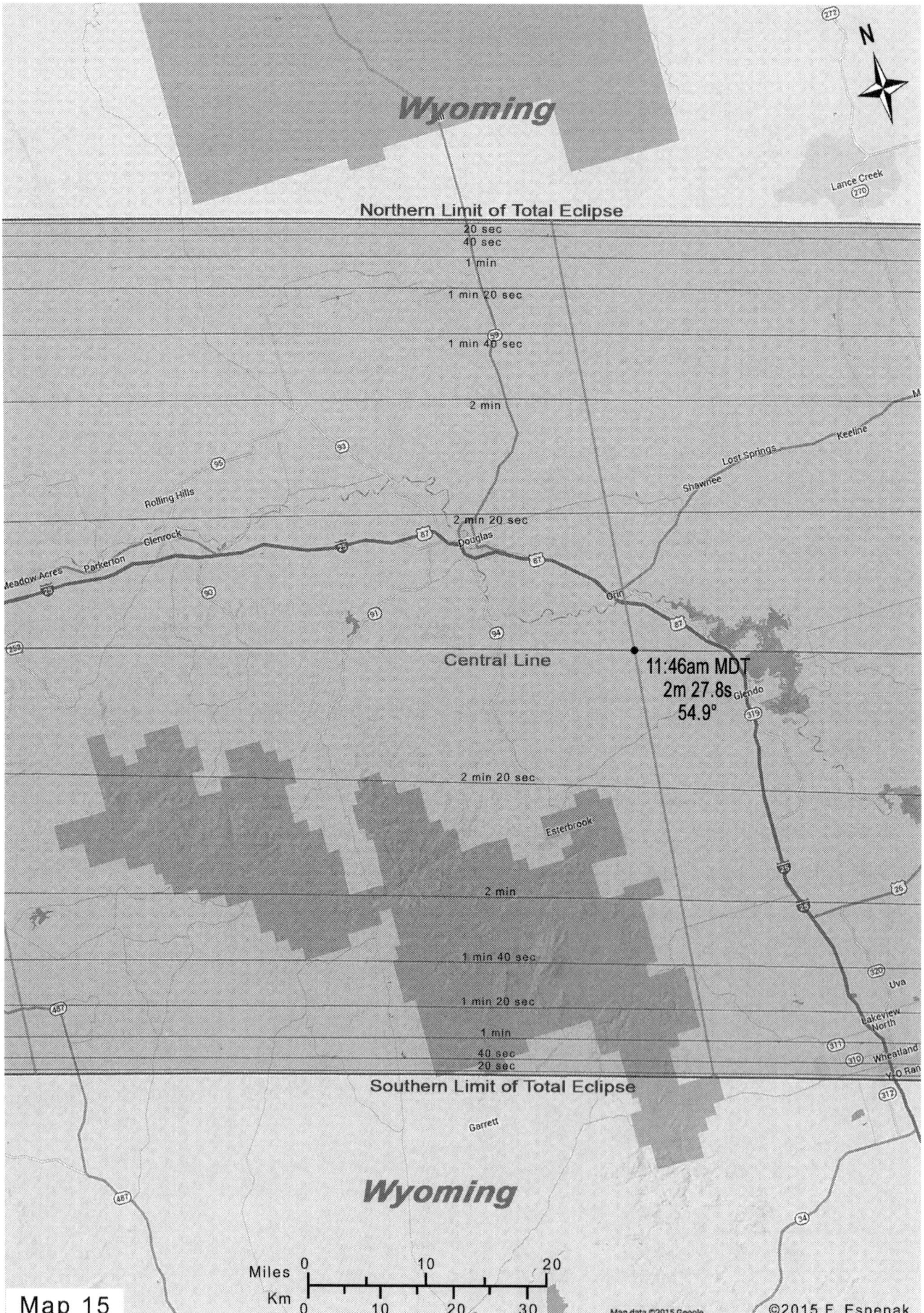
Wyoming
Northern Limit of Total Eclipse
20 sec
40 sec
1 min
1 min 20 sec
1 min 40 sec
2 min
2 min 20 sec
Rolling Hills
Glenrock
Parkerton
Meadow Acres
Douglas
Lost Springs
Keeline
Shawnee
Orin
Lance Creek
Central Line
11:46am MDT
2m 27.8s
54.9°
Glendo
2 min 20 sec
Esterbrook
2 min
1 min 40 sec
1 min 20 sec
1 min
40 sec
20 sec
Southern Limit of Total Eclipse
Uva
Lakeview North
Wheatland
Garrett
Wyoming
Miles 0 10 20
Km 0 10 20 30
Map 15
Map data ©2015 Google
©2015 F. Espenak

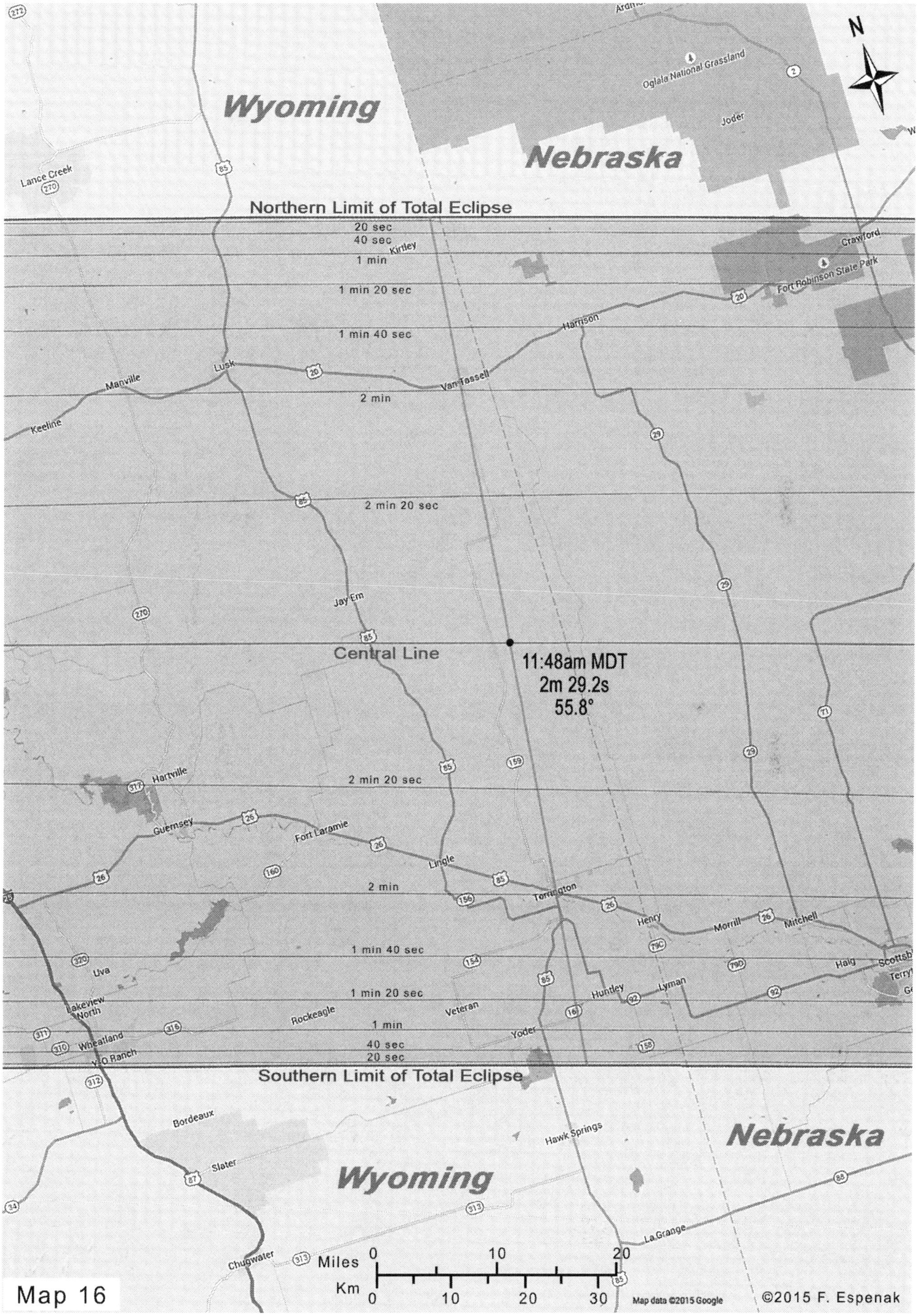
N
Wyoming
Nebraska
Oglala National Grassland
Joder
Lance Creek
Northern Limit of Total Eclipse
20 sec
40 sec
Kirtley
1 min
1 min 20 sec
Crawford
Fort Robinson State Park
Harrison
1 min 40 sec
Lusk
Manville
Van Tassell
2 min
Keeline
2 min 20 sec
Jay Em
Central Line
11:48am MDT
2m 29.2s
55.8°
Hartville
2 min 20 sec
Guernsey
Fort Laramie
Lingle
2 min
Torrington
Henry
Morrill
Mitchell
1 min 40 sec
Uva
Haig
Huntley
Lyman
1 min 20 sec
Lakeview North
Veteran
Rockeagle
1 min
Wheatland
Yoder
40 sec
20 sec
Y-O Ranch
Southern Limit of Total Eclipse
Bordeaux
Hawk Springs
Nebraska
Slater
Wyoming
La Grange
Chugwater
Miles
0
10
20
Km
0
10
20
30
Map data ©2015 Google
©2015 F. Espenak

Map 16

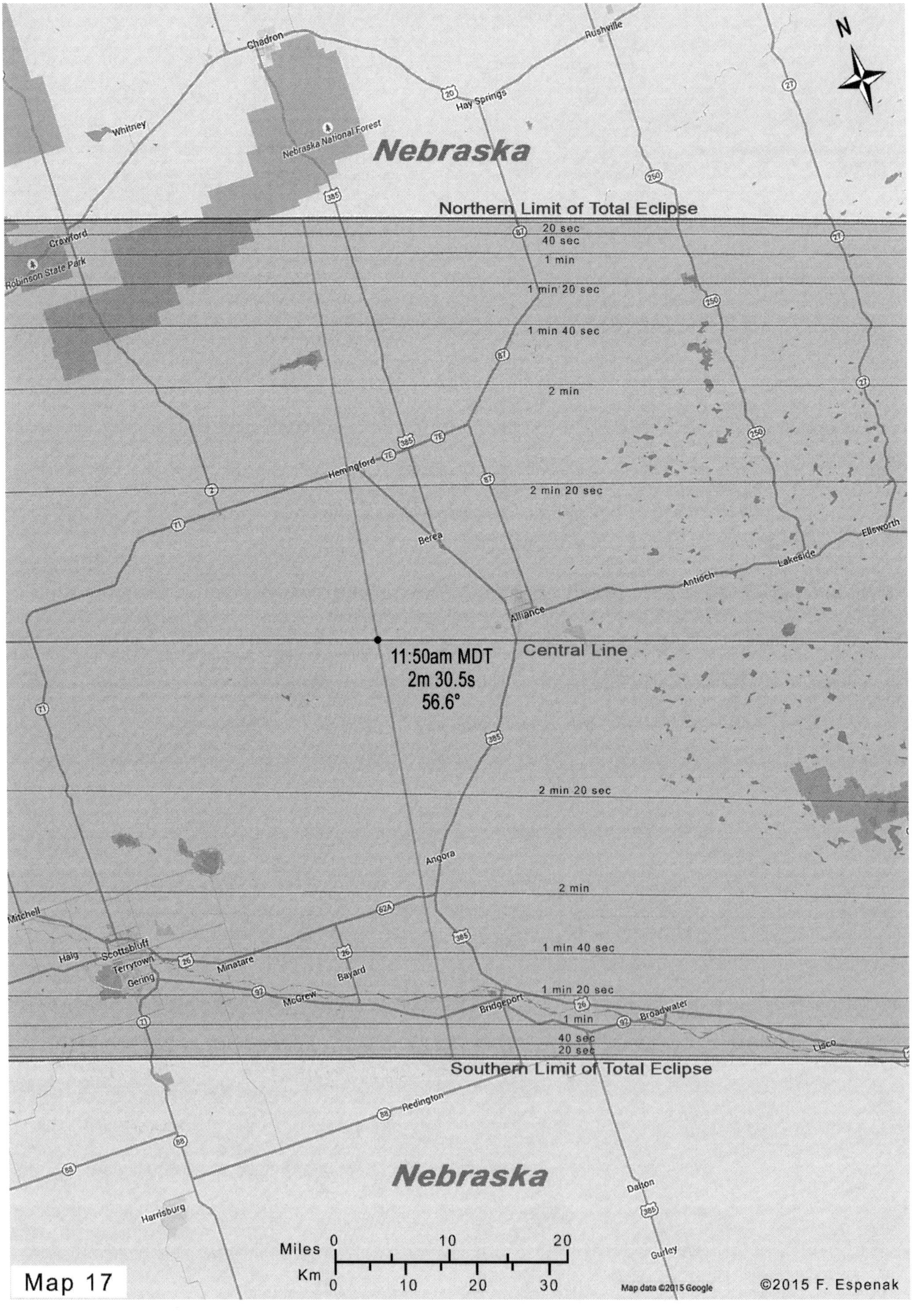
Nebraska
Chadron
Rushville
Hay Springs
Whitney
Nebraska National Forest
Northern Limit of Total Eclipse
20 sec
40 sec
1 min
1 min 20 sec
1 min 40 sec
2 min
2 min 20 sec
Crawford
Robinson State Park
Hemingford
Berea
Alliance
Antioch
Lakeside
Ellsworth
11:50am MDT
2m 30.5s
56.6°
Central Line
2 min 20 sec
Angora
2 min
Mitchell
Scottsbluff
Haig
Terrytown
Gering
Minatare
Bayard
McGrew
Bridgeport
Broadwater
Lisco
1 min 40 sec
1 min 20 sec
1 min
40 sec
20 sec
Southern Limit of Total Eclipse
Redington
Nebraska
Dalton
Harrisburg
Gurley
Miles 0 10 20
Km 0 10 20 30
Map 17
Map data ©2015 Google
©2015 F. Espenak

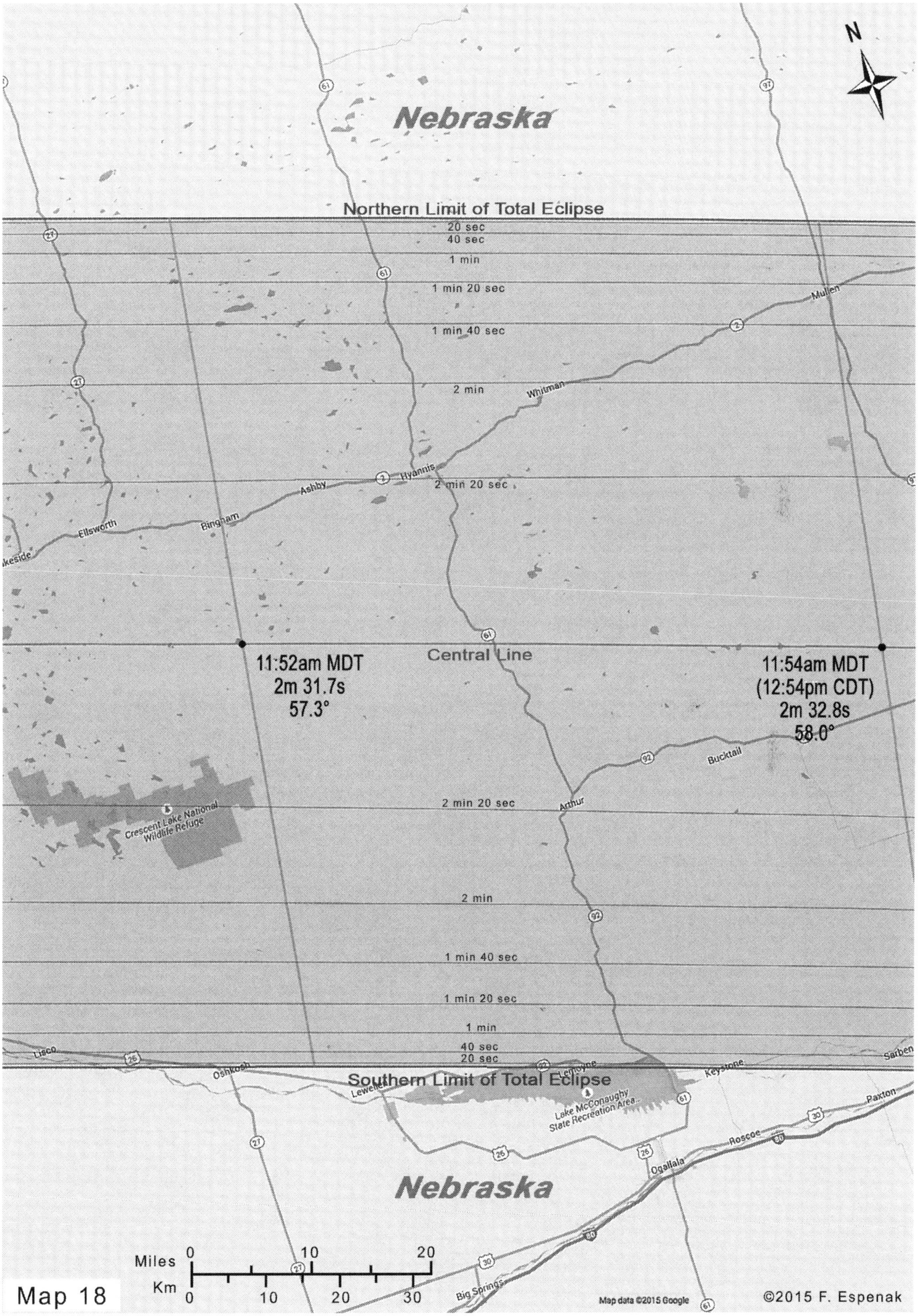

N
Nebraska
Northern Limit of Total Eclipse
20 sec
40 sec
1 min
1 min 20 sec
1 min 40 sec
2 min
2 min 20 sec
Mullen
Whitman
Hyannis
Ashby
Bingham
Ellsworth
Central Line
11:52am MDT
2m 31.7s
57.3°
11:54am MDT
(12:54pm CDT)
2m 32.8s
58.0°
Bucktail
Arthur
Crescent Lake National Wildlife Refuge
2 min 20 sec
2 min
1 min 40 sec
1 min 20 sec
1 min
40 sec
20 sec
Lisco
Oshkosh
Lemoyne
Keystone
Sarben
Southern Limit of Total Eclipse
Lewellen
Lake McConaughy State Recreation Area
Paxton
Roscoe
Ogallala
Nebraska
Big Springs
Miles 0 10 20
Km 0 10 20 30
Map 18
Map data ©2015 Google
©2015 F. Espenak

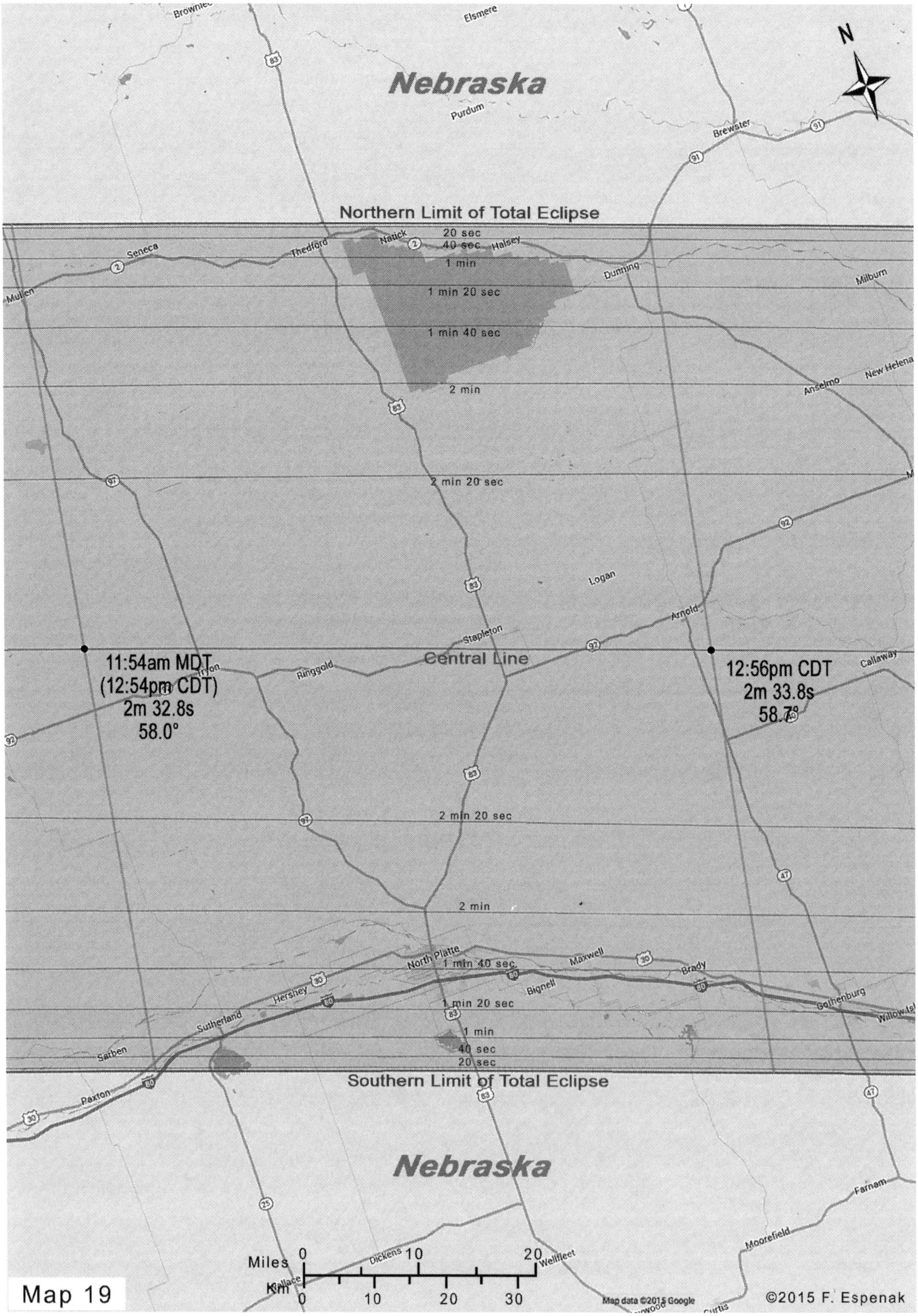
Nebraska
Brownlee
Elsmere
Purdum
Brewster
Northern Limit of Total Eclipse
20 sec
40 sec
1 min
1 min 20 sec
1 min 40 sec
2 min
2 min 20 sec
Seneca
Thedford
Natick
Halsey
Dunning
Milburn
Mullen
Anselmo
New Helena
Logan
Arnold
Stapleton
Central Line
Ringgold
Tryon
Callaway
11:54am MDT
(12:54pm CDT)
2m 32.8s
58.0°
12:56pm CDT
2m 33.8s
58.7°
2 min 20 sec
2 min
North Platte
1 min 40 sec
Maxwell
Brady
Hershey
Bignell
1 min 20 sec
Gothenburg
Sutherland
1 min
40 sec
20 sec
Sarben
Southern Limit of Total Eclipse
Paxton
Nebraska
Farnam
Moorefield
Dickens
Wellfleet
Wallace
Curtis
Miles 0 10 20
Km 0 10 20 30
Map 19
Map data ©2015 Google
©2015 F. Espenak

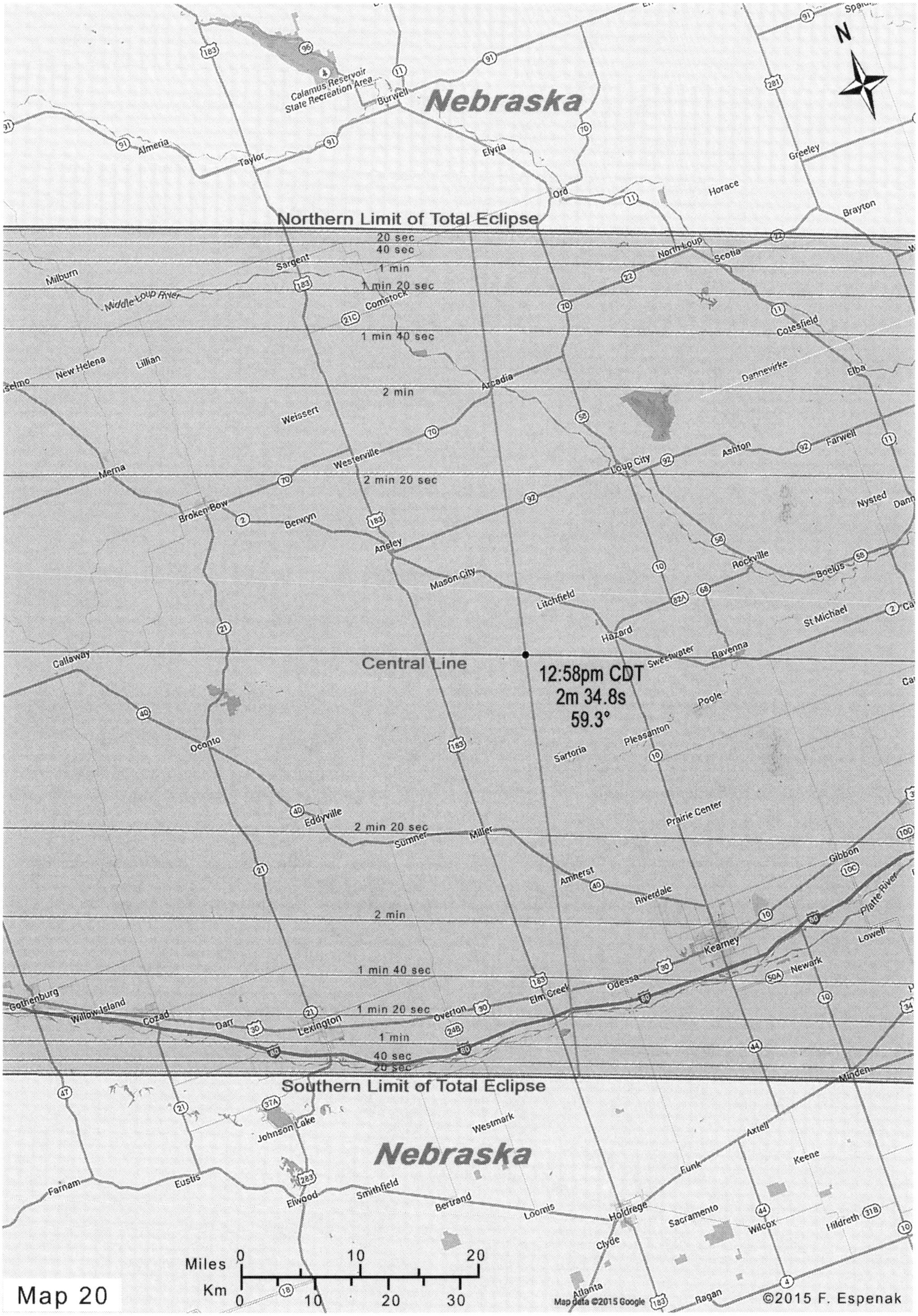
Nebraska
Northern Limit of Total Eclipse
20 sec
40 sec
1 min
1 min 20 sec
1 min 40 sec
2 min
2 min 20 sec
Central Line
12:58pm CDT
2m 34.8s
59.3°
2 min 20 sec
2 min
1 min 40 sec
1 min 20 sec
1 min
40 sec
20 sec
Southern Limit of Total Eclipse
Nebraska
Calamus Reservoir State Recreation Area
Burwell
Almeria
Taylor
Elyria
Ord
Horace
Greeley
Brayton
Sargent
Milburn
Middle Loup River
Comstock
North Loup
Scotia
Cotesfield
Dannevirke
Elba
New Helena
Lillian
Arcadia
Weissert
Westerville
Ashton
Farwell
Loup City
Merna
Broken Bow
Berwyn
Ansley
Nysted
Rockville
Boelus
Mason City
Litchfield
Hazard
St Michael
Callaway
Sweetwater
Ravenna
Poole
Oconto
Pleasanton
Sartoria
Eddyville
Prairie Center
Sumner
Miller
Gibbon
Amherst
Riverdale
Platte River
Kearney
Lowell
Newark
Gothenburg
Willow Island
Cozad
Darr
Lexington
Overton
Elm Creek
Odessa
Minden
Johnson Lake
Westmark
Axtell
Keene
Funk
Farnam
Eustis
Elwood
Smithfield
Bertrand
Loomis
Holdrege
Sacramento
Wilcox
Hildreth
Clyde
Atlanta
Ragan
Miles 0 10 20
Km 0 10 20 30
Map 20
Map data ©2015 Google
©2015 F. Espenak

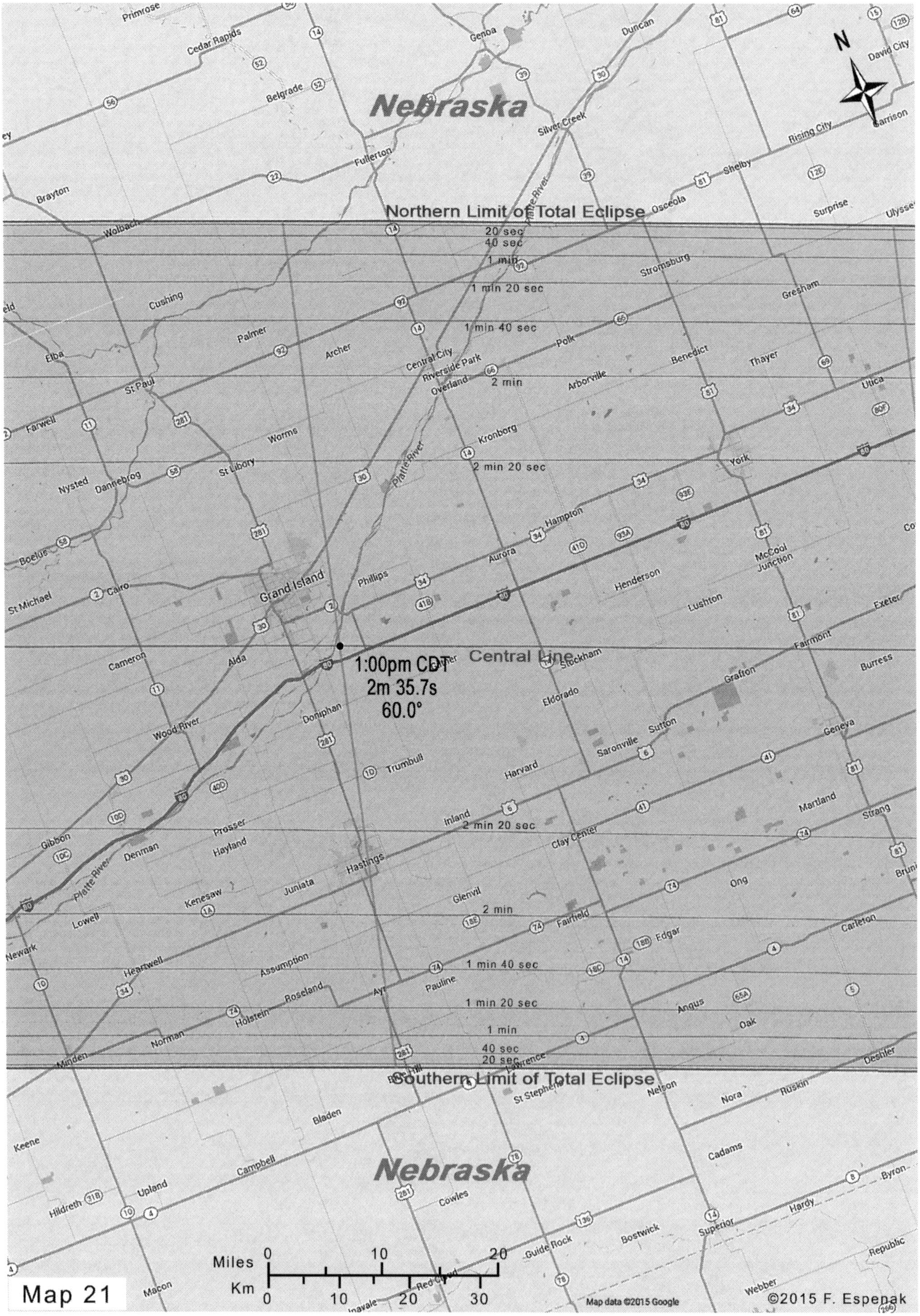
Nebraska
Northern Limit of Total Eclipse
20 sec
40 sec
1 min
1 min 20 sec
1 min 40 sec
2 min
2 min 20 sec
Central Line
1:00pm CDT
2m 35.7s
60.0°
2 min 20 sec
2 min
1 min 40 sec
1 min 20 sec
1 min
40 sec
20 sec
Southern Limit of Total Eclipse
Nebraska
N
Primrose
Cedar Rapids
Belgrade
Genoa
Duncan
David City
Silver Creek
Fullerton
Rising City
Garrison
Shelby
Brayton
Osceola
Surprise
Ulysses
Wolbach
Stromsburg
Cushing
Gresham
Palmer
Polk
Elba
Archer
Central City
Riverside Park
Overland
Benedict
Thayer
Arborville
St Paul
Utica
Farwell
Worms
Kronborg
Platte River
York
Nysted
Dannebrog
St Libory
Hampton
Boelus
Aurora
McCool Junction
St Michael
Cairo
Grand Island
Phillips
Henderson
Lushton
Exeter
Fairmont
Cameron
Alda
Stockham
Burress
Grafton
Eldorado
Doniphan
Wood River
Sutton
Saronville
Geneva
Trumbull
Harvard
Martland
Inland
Strang
Gibbon
Prosser
Hayland
Clay Center
Denman
Hastings
Ong
Kenesaw
Juniata
Glenvil
Lowell
Fairfield
Carleton
Newark
Edgar
Heartwell
Assumption
Pauline
Roseland
Ayr
Angus
Holstein
Oak
Norman
Minden
Lawrence
Deshler
St Stephen
Nelson
Nora
Ruskin
Bladen
Keene
Cadams
Campbell
Upland
Hildreth
Cowles
Byron
Hardy
Superior
Bostwick
Guide Rock
Republic
Red Cloud
Inavale
Webber
Macon
Miles 0 10 20
Km 0 10 20 30
Map 21
Map data ©2015 Google
©2015 F. Espenak

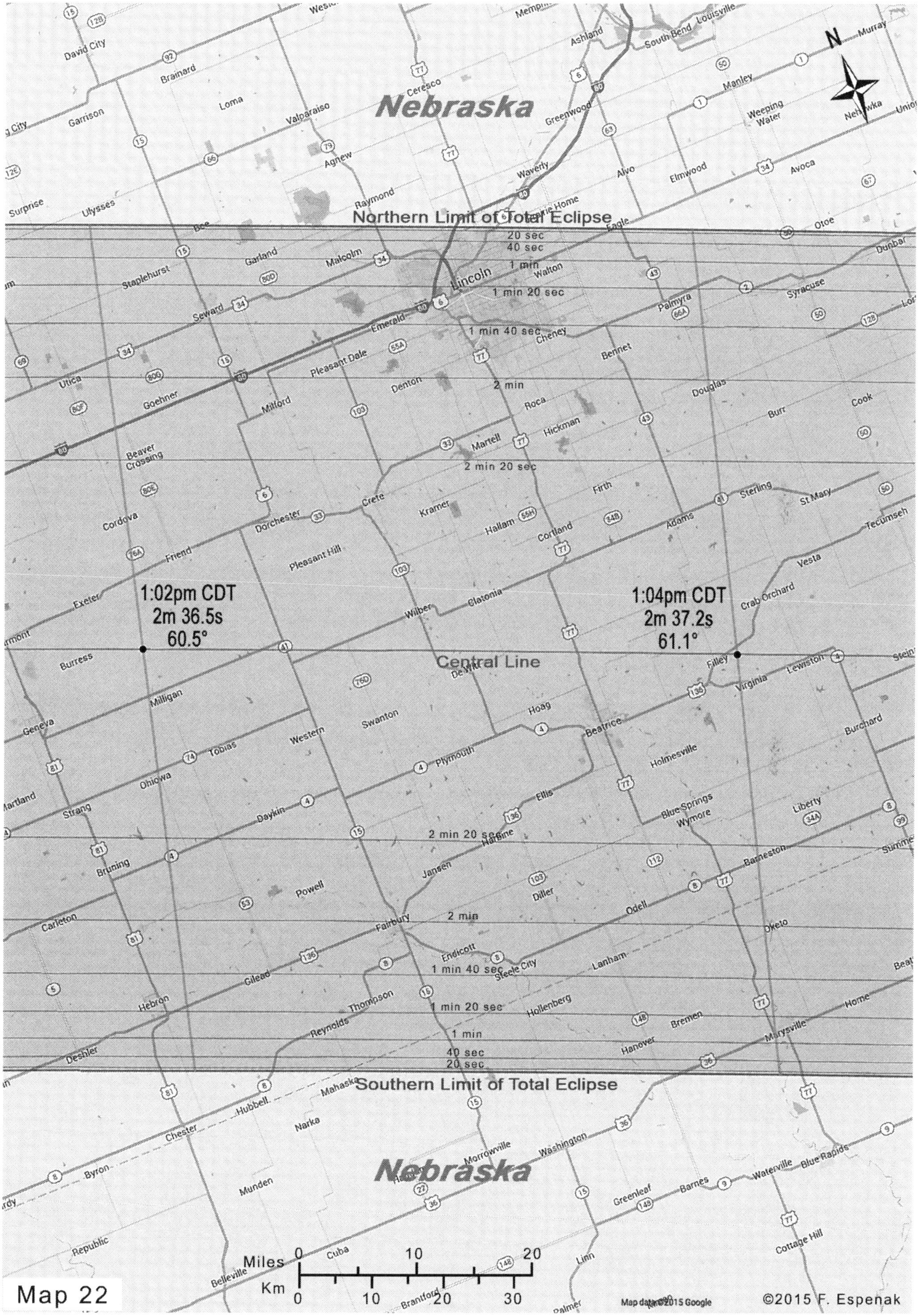
Nebraska
N
Northern Limit of Total Eclipse
20 sec
40 sec
1 min
1 min 20 sec
1 min 40 sec
2 min
2 min 20 sec
1:02pm CDT
2m 36.5s
60.5°
1:04pm CDT
2m 37.2s
61.1°
Central Line
2 min 20 sec
2 min
1 min 40 sec
1 min 20 sec
1 min
40 sec
20 sec
Southern Limit of Total Eclipse
Nebraska
Lincoln
Beatrice
Fairbury
Crete
Wilber
Seward
Hebron
Marysville
Miles 0 10 20
Km 0 10 20 30
Map 22
Map data ©2015 Google
©2015 F. Espenak

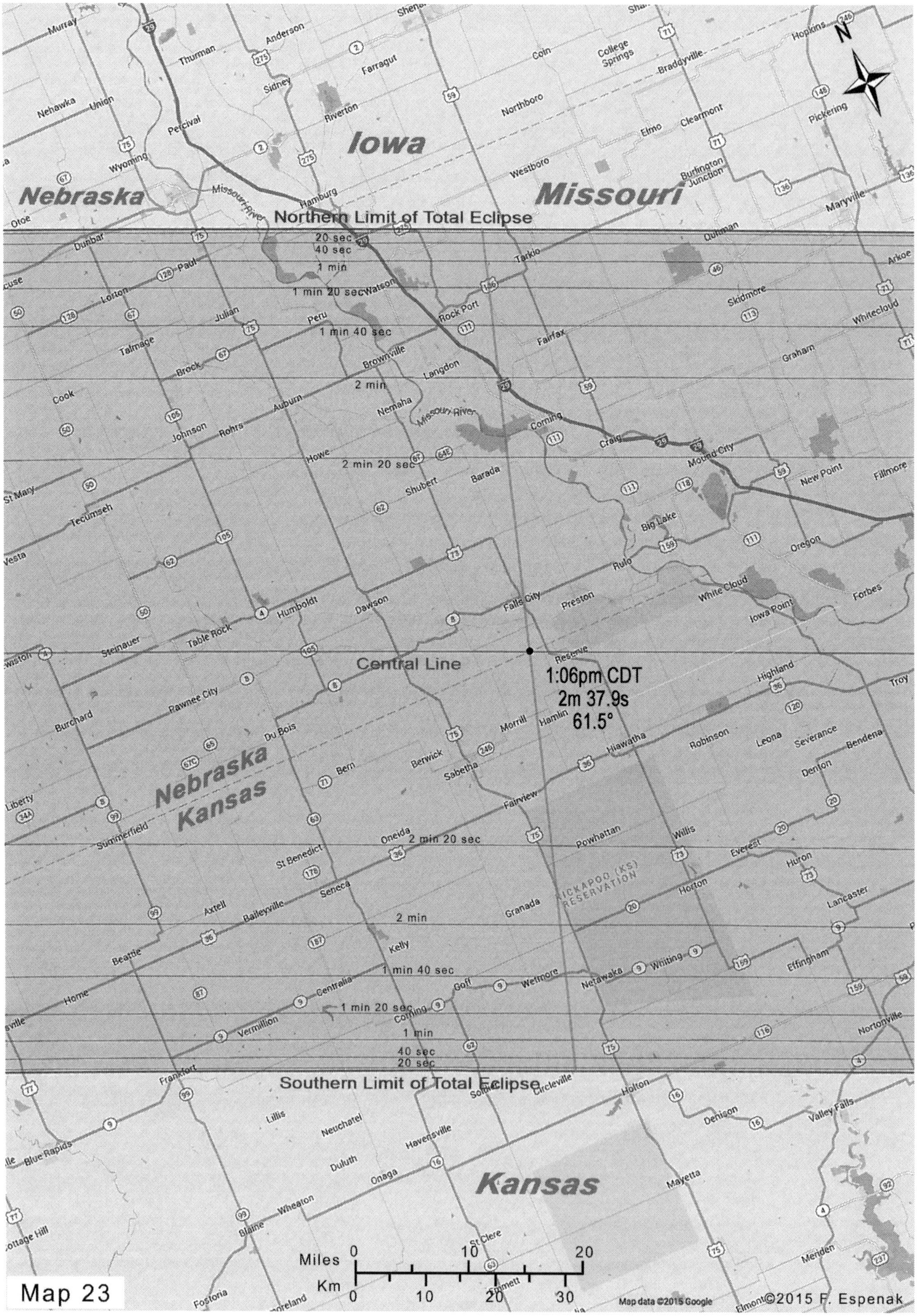
Iowa
Nebraska
Missouri
Northern Limit of Total Eclipse
20 sec
40 sec
1 min
1 min 20 sec
1 min 40 sec
2 min
2 min 20 sec
Central Line
1:06pm CDT
2m 37.9s
61.5°
Nebraska
Kansas
2 min 20 sec
2 min
1 min 40 sec
1 min 20 sec
1 min
40 sec
20 sec
Southern Limit of Total Eclipse
Kansas
Miles 0 10 20
Km 0 10 20 30
Map 23
Map data ©2015 Google
©2015 F. Espenak

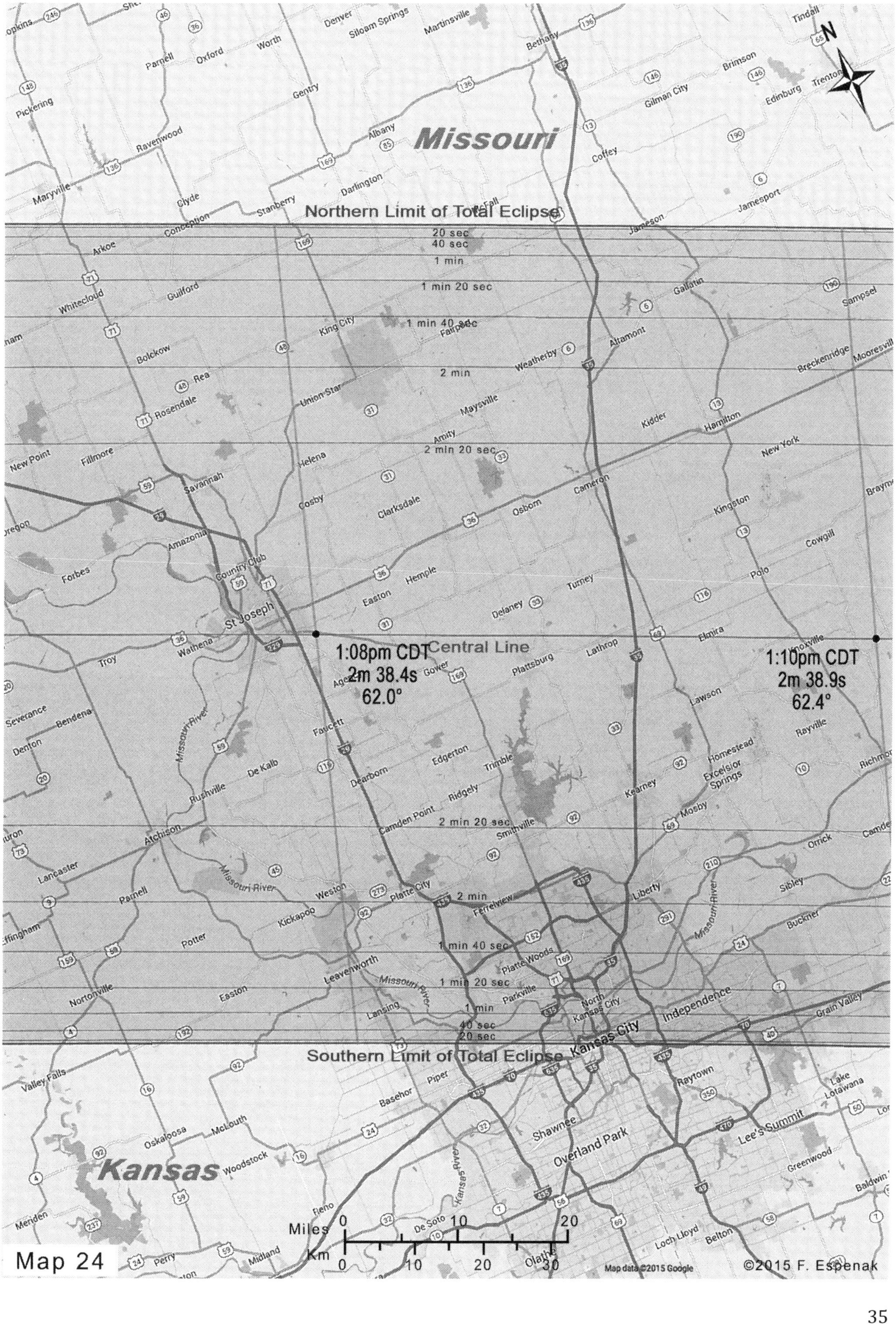
Missouri
Kansas
Northern Limit of Total Eclipse
Southern Limit of Total Eclipse
Central Line
20 sec
40 sec
1 min
1 min 20 sec
1 min 40 sec
2 min
2 min 20 sec
1:08pm CDT
2m 38.4s
62.0°
1:10pm CDT
2m 38.9s
62.4°
St Joseph
Kansas City
Independence
Liberty
Lee's Summit
Overland Park
Shawnee
Leavenworth
Atchison
Cameron
Maryville
Miles 0 10 20
Km 0 10 20 30
Map data ©2015 Google
©2015 F. Espenak
Map 24

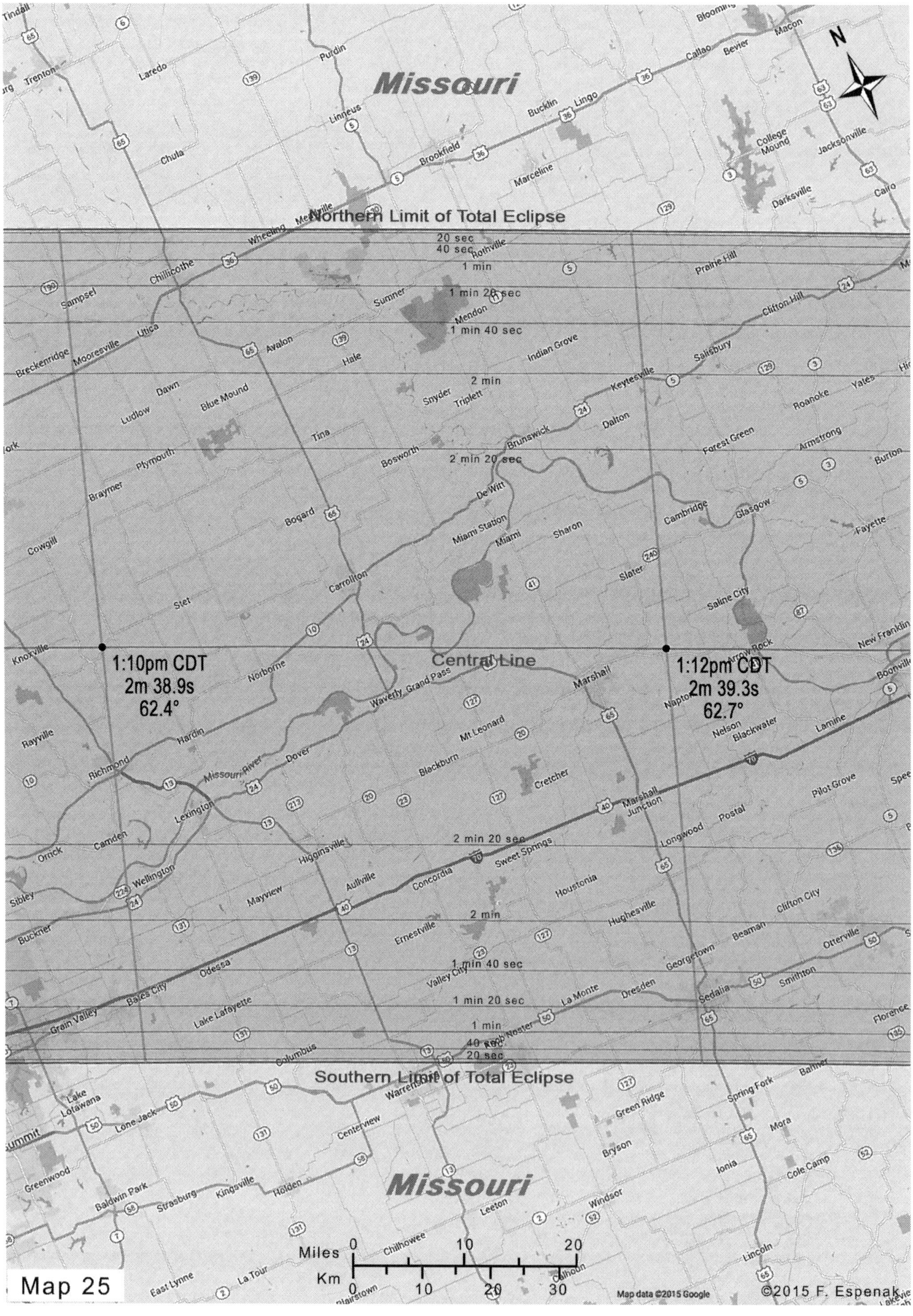
Missouri
N
Northern Limit of Total Eclipse
20 sec
40 sec
1 min
1 min 20 sec
1 min 40 sec
2 min
2 min 20 sec
Central Line
1:10pm CDT
2m 38.9s
62.4°
1:12pm CDT
2m 39.3s
62.7°
2 min 20 sec
2 min
1 min 40 sec
1 min 20 sec
1 min
40 sec
20 sec
Southern Limit of Total Eclipse
Missouri
Trenton
Laredo
Purdin
Linneus
Brookfield
Marceline
Bucklin
Lingo
Callao
Bevier
Macon
College Mound
Jacksonville
Darksville
Cairo
Chula
Chillicothe
Wheeling
Rothville
Sumner
Mendon
Prairie Hill
Clifton Hill
Sampsel
Utica
Mooresville
Breckenridge
Avalon
Hale
Indian Grove
Salisbury
Dawn
Blue Mound
Ludlow
Snyder
Triplett
Keytesville
Dalton
Roanoke
Yates
Tina
Brunswick
Forest Green
Armstrong
Burton
Plymouth
Braymer
Bosworth
De Witt
Glasgow
Cambridge
Bogard
Miami Station
Miami
Sharon
Fayette
Cowgill
Carrollton
Slater
Saline City
Stet
Knoxville
Norborne
Waverly
Grand Pass
Marshall
Arrow Rock
New Franklin
Boonville
Napton
Nelson
Blackwater
Lamine
Rayville
Hardin
Mt Leonard
Richmond
Dover
Missouri River
Blackburn
Cretcher
Pilot Grove
Lexington
Marshall Junction
Postal
Longwood
Orrick
Camden
Sweet Springs
Wellington
Higginsville
Concordia
Houstonia
Sibley
Mayview
Aullville
Clifton City
Hughesville
Buckner
Ernestville
Beaman
Otterville
Odessa
Valley City
Georgetown
Smithton
Bates City
La Monte
Dresden
Sedalia
Grain Valley
Lake Lafayette
Knob Noster
Florence
Columbus
Warrensburg
Spring Fork
Lake Lotawana
Lone Jack
Green Ridge
Summit
Centerview
Mora
Bryson
Greenwood
Cole Camp
Ionia
Holden
Kingsville
Strasburg
Baldwin Park
Leeton
Windsor
Chilhowee
Lincoln
East Lynne
La Tour
Blairstown
Calhoun
Miles 0 10 20
Km 0 10 20 30
Map 25
Map data ©2015 Google
©2015 F. Espenak

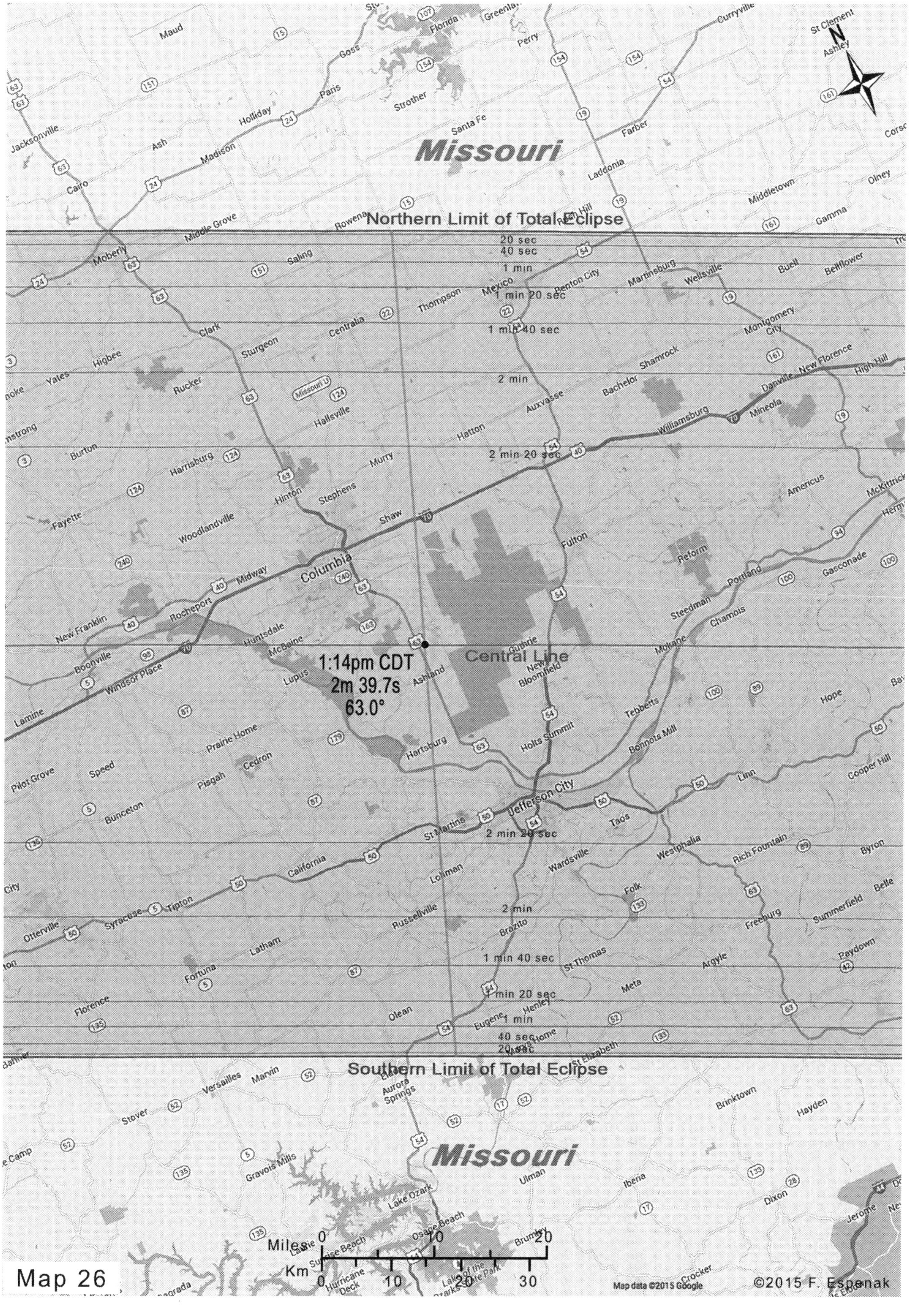
Missouri
Northern Limit of Total Eclipse
20 sec
40 sec
1 min
1 min 20 sec
1 min 40 sec
2 min
2 min 20 sec
1:14pm CDT
2m 39.7s
63.0°
Central Line
2 min 20 sec
2 min
1 min 40 sec
1 min 20 sec
1 min
40 sec
20 sec
Southern Limit of Total Eclipse
Missouri
Columbia
Jefferson City
Mexico
Fulton
Moberly
Boonville
California
Tipton
Ashland
Hartsburg
Holts Summit
New Bloomfield
Miles 0 10 20
Km 0 10 20 30
Map 26
Map data ©2015 Google
©2015 F. Espenak

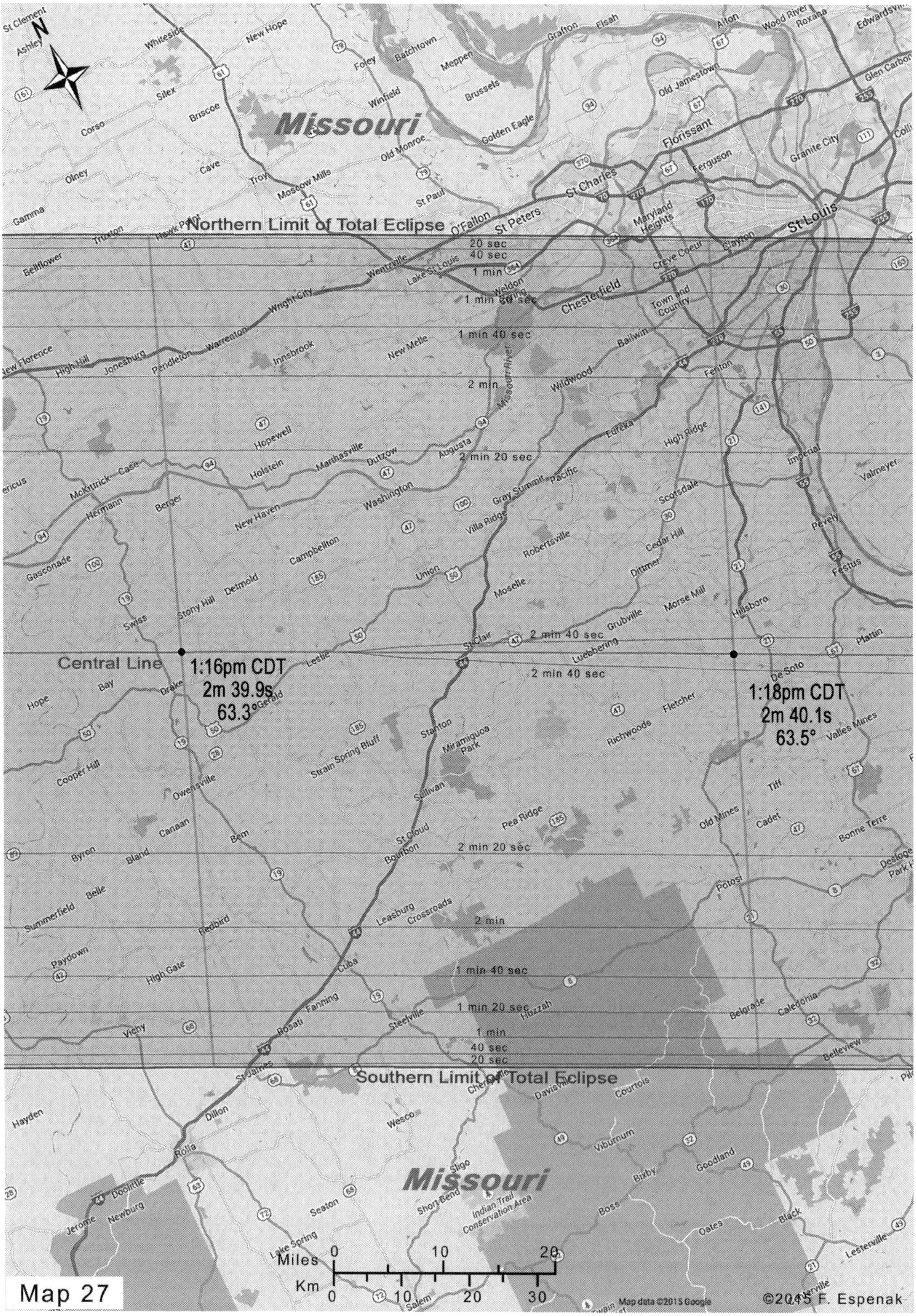
N
Missouri
Northern Limit of Total Eclipse
20 sec
40 sec
1 min
1 min 20 sec
1 min 40 sec
2 min
2 min 20 sec
2 min 40 sec
Central Line
1:16pm CDT
2m 39.9s
63.3°
1:18pm CDT
2m 40.1s
63.5°
2 min 40 sec
2 min 20 sec
2 min
1 min 40 sec
1 min 20 sec
1 min
40 sec
20 sec
Southern Limit of Total Eclipse
Missouri
St Clement
Ashley
Whiteside
New Hope
Foley
Batchtown
Meppen
Grafton
Elsah
Alton
Wood River
Roxana
Edwardsville
Glen Carbon
Silex
Winfield
Brussels
Briscoe
Corso
Golden Eagle
Old Jamestown
Florissant
Granite City
Olney
Cave
Old Monroe
Troy
Moscow Mills
St Paul
St Charles
Ferguson
Maryland Heights
Gamma
Truxton
Hawk Point
O'Fallon
St Peters
St Louis
Clayton
Creve Coeur
Bellflower
Wentzville
Lake St Louis
Weldon Spring
Chesterfield
Town and Country
Wright City
New Florence
High Hill
Jonesburg
Pendleton
Warrenton
Innsbrook
New Melle
Missouri River
Ballwin
Wildwood
Fenton
Hopewell
Augusta
Eureka
High Ridge
Imperial
Valmeyer
Holstein
Marthasville
Dutzow
McKittrick
Case
Hermann
Berger
New Haven
Washington
Gray Summit
Pacific
Scotsdale
Pevely
Villa Ridge
Robertsville
Cedar Hill
Festus
Gasconade
Campbellton
Union
Dittmer
Detmold
Stony Hill
Moselle
Morse Mill
Swiss
Hillsboro
Grubville
St Clair
Plattin
Luebbering
Leslie
De Soto
Hope
Bay
Drake
Gerald
Fletcher
Richwoods
Valles Mines
Stanton
Miramiguoa Park
Strain Spring Bluff
Cooper Hill
Owensville
Sullivan
Tiff
Old Mines
Canaan
Pea Ridge
Cadet
Bonne Terre
Byron
Bland
Bem
St Cloud
Bourbon
Desloge
Park Hills
Belle
Summerfield
Potosi
Redbird
Leasburg
Crossroads
Paydown
High Gate
Cuba
Fanning
Steelville
Huzzah
Belgrade
Caledonia
Vichy
Rosati
Belleview
St James
Cherryville
Davisville
Courtois
Hayden
Dillon
Wesco
Viburnum
Rolla
Sligo
Goodland
Doolittle
Newburg
Short Bend
Indian Trail Conservation Area
Bixby
Boss
Jerome
Seaton
Oates
Black
Lesterville
Lake Spring
Salem
Miles
0
10
20
Km
0
10
20
30
Map 27
Map data ©2015 Google
©2015 F. Espenak

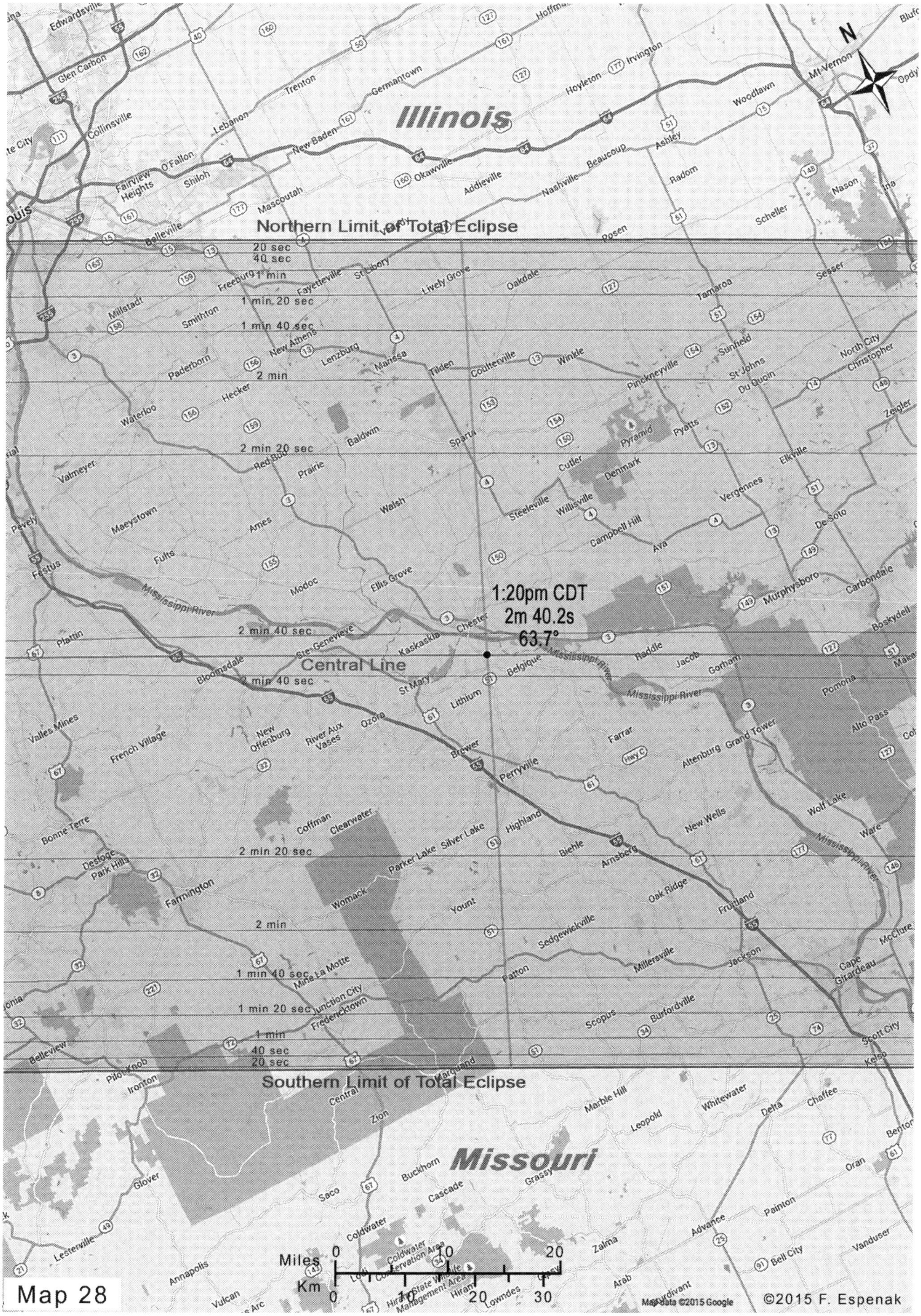
Illinois
Missouri
Northern Limit of Total Eclipse
20 sec
40 sec
1 min
1 min 20 sec
1 min 40 sec
2 min
2 min 20 sec
2 min 40 sec
Central Line
2 min 40 sec
2 min 20 sec
2 min
1 min 40 sec
1 min 20 sec
1 min
40 sec
20 sec
Southern Limit of Total Eclipse
1:20pm CDT
2m 40.2s
63.7°
Mississippi River
N
Miles 0 10 20
Km 0 10 20 30
Map 28
Map data ©2015 Google
©2015 F. Espenak
Edwardsville
Glen Carbon
Collinsville
Trenton
Germantown
Lebanon
New Baden
O'Fallon
Shiloh
Fairview Heights
Mascoutah
Belleville
Okawville
Addieville
Nashville
Beaucoup
Ashley
Hoyleton
Irvington
Woodlawn
Mt Vernon
Radom
Scheller
Nason
Rosen
Freeburg
Fayetteville
St Libory
Lively Grove
Oakdale
Tamaroa
Sesser
Millstadt
Smithton
New Athens
Lenzburg
Marissa
Tilden
Coulterville
Winkle
Pinckneyville
Sunfield
St Johns
Du Quoin
North City
Christopher
Zeigler
Paderborn
Hecker
Waterloo
Baldwin
Sparta
Pyramid
Pyatts
Red Bud
Prairie
Cutler
Denmark
Elkville
Valmeyer
Walsh
Steeleville
Willisville
Vergennes
De Soto
Maeystown
Ames
Campbell Hill
Ava
Pevely
Fults
Festus
Modoc
Ellis Grove
Murphysboro
Carbondale
Chester
Plattin
Ste Genevieve
Kaskaskia
Belgique
Raddle
Jacob
Gorham
Boskydell
Bloomsdale
St Mary
Lithium
Pomona
Valles Mines
New Offenburg
River Aux Vases
Ozora
Farrar
Grand Tower
Altenburg
Alto Pass
French Village
Brewer
Perryville
Bonne Terre
Coffman
Clearwater
Highland
New Wells
Wolf Lake
Desloge
Park Hills
Parker Lake
Silver Lake
Biehle
Arnsberg
Ware
Farmington
Womack
Yount
Oak Ridge
Fruitland
Sedgewickville
Millersville
Jackson
Cape Girardeau
McClure
Mine La Motte
Patton
Junction City
Fredericktown
Scopus
Burfordville
Scott City
Kelso
Belleview
Pilot Knob
Ironton
Marquand
Central
Zion
Marble Hill
Whitewater
Leopold
Delta
Chaffee
Benton
Glover
Buckhorn
Saco
Cascade
Grassy
Oran
Lesterville
Coldwater
Zalma
Advance
Painton
Vanduser
Annapolis
Vulcan
Lowndes
Arab
Bell City
Sturdivant

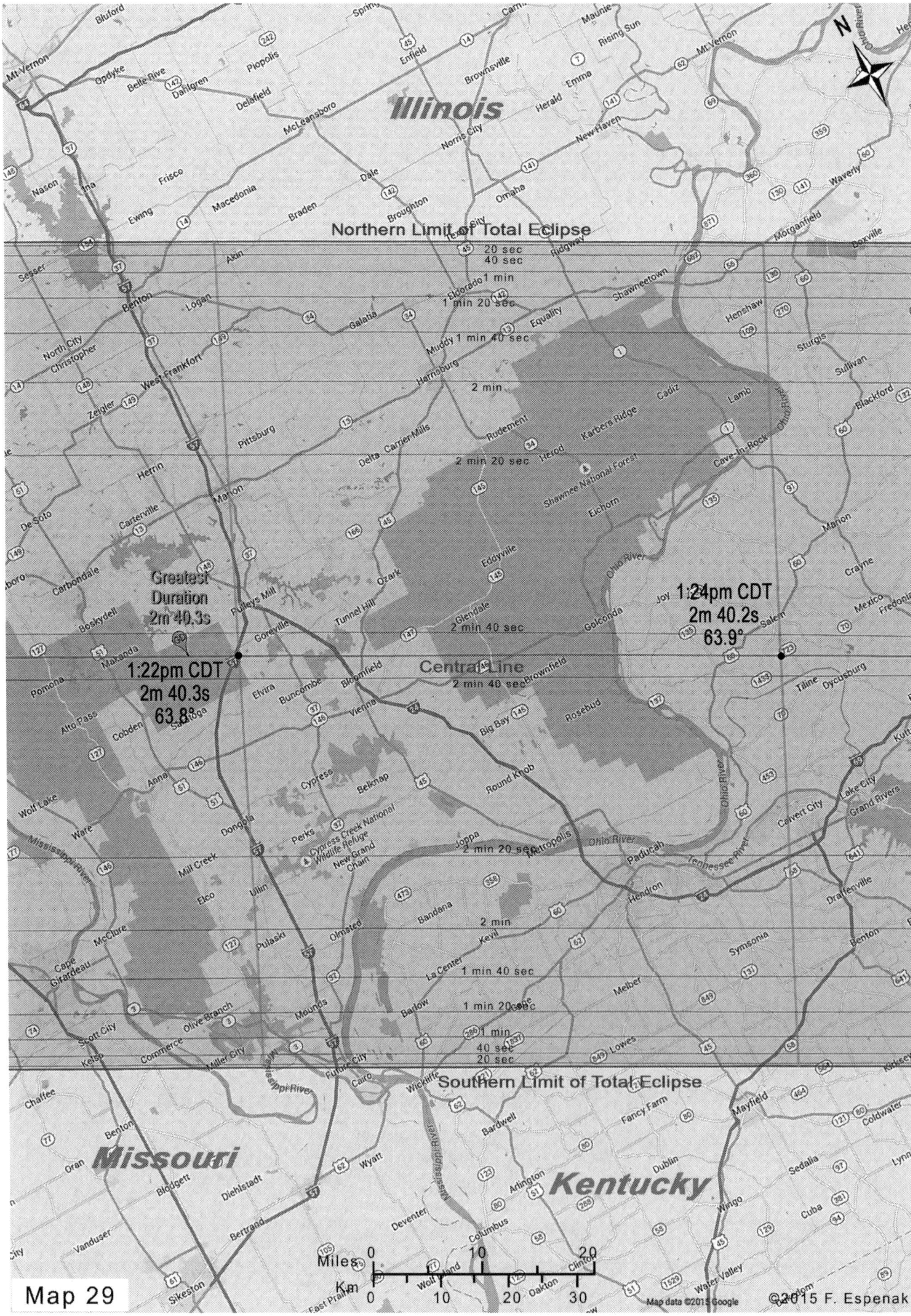
Illinois
Missouri
Kentucky
Northern Limit of Total Eclipse
Southern Limit of Total Eclipse
Central Line
20 sec
40 sec
1 min
1 min 20 sec
1 min 40 sec
2 min
2 min 20 sec
2 min 40 sec
Greatest Duration 2m 40.3s
1:22pm CDT
2m 40.3s
63.8°
1:24pm CDT
2m 40.2s
63.9°
Shawnee National Forest
Cypress Creek National Wildlife Refuge
Ohio River
Mississippi River
Tennessee River
Mt Vernon
Benton
Marion
Carbondale
Cape Girardeau
Paducah
Metropolis
Golconda
Eddyville
Harrisburg
Shawneetown
Cave-In-Rock
Cairo
Sikeston
Mayfield
Miles 0 10 20
Km 0 10 20 30
Map data ©2015 Google
©2015 F. Espenak
Map 29

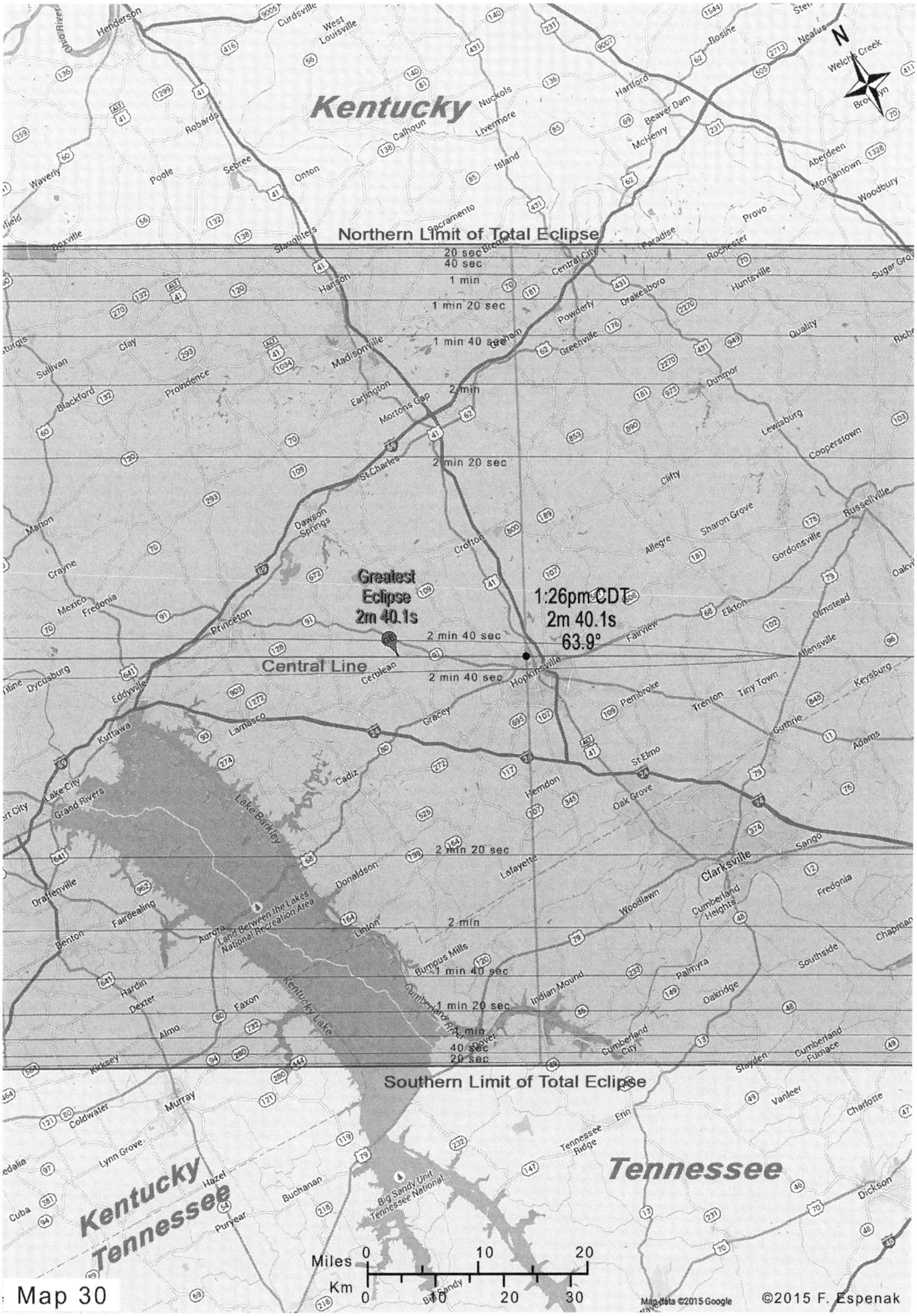
Kentucky
Northern Limit of Total Eclipse
20 sec
40 sec
1 min
1 min 20 sec
1 min 40 sec
2 min
2 min 20 sec
Greatest Eclipse 2m 40.1s
1:26pm CDT
2m 40.1s
63.9°
2 min 40 sec
Central Line
2 min 40 sec
2 min 20 sec
2 min
1 min 40 sec
1 min 20 sec
1 min
40 sec
20 sec
Southern Limit of Total Eclipse
Tennessee
Kentucky
Tennessee
Hopkinsville
Clarksville
Princeton
Madisonville
Lake Barkley
Kentucky Lake
Land Between the Lakes National Recreation Area
Murray
N
Miles 0 10 20
Km 0 10 20 30
Map 30
Map data ©2015 Google
©2015 F. Espenak

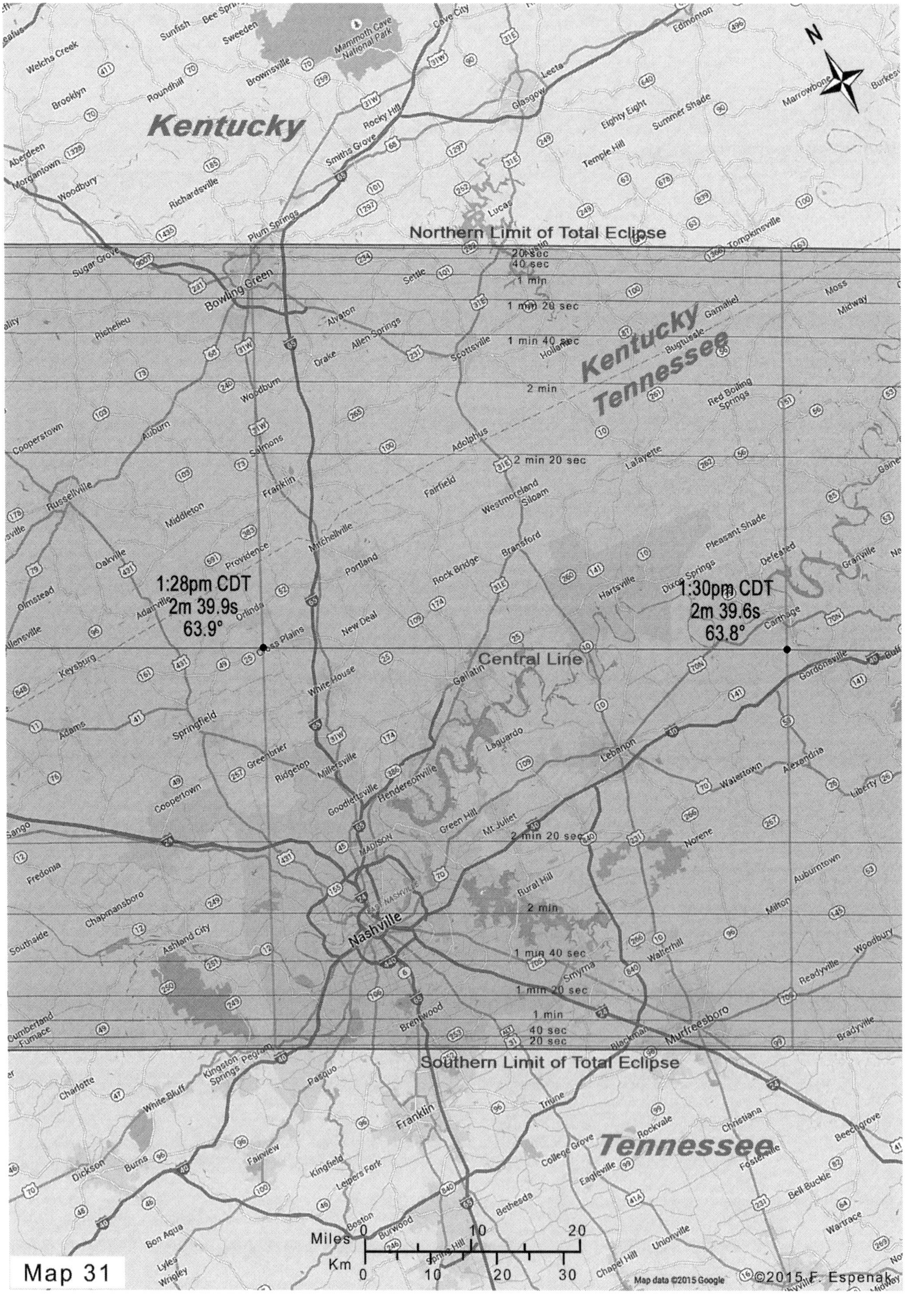

Kentucky
Tennessee
Northern Limit of Total Eclipse
Southern Limit of Total Eclipse
Central Line
1:28pm CDT
2m 39.9s
63.9°
1:30pm CDT
2m 39.6s
63.8°
20 sec
40 sec
1 min
1 min 20 sec
1 min 40 sec
2 min
2 min 20 sec
Bowling Green
Nashville
Murfreesboro
Franklin
Miles 0 10 20
Km 0 10 20 30
Map 31
Map data ©2015 Google
©2015 F. Espenak

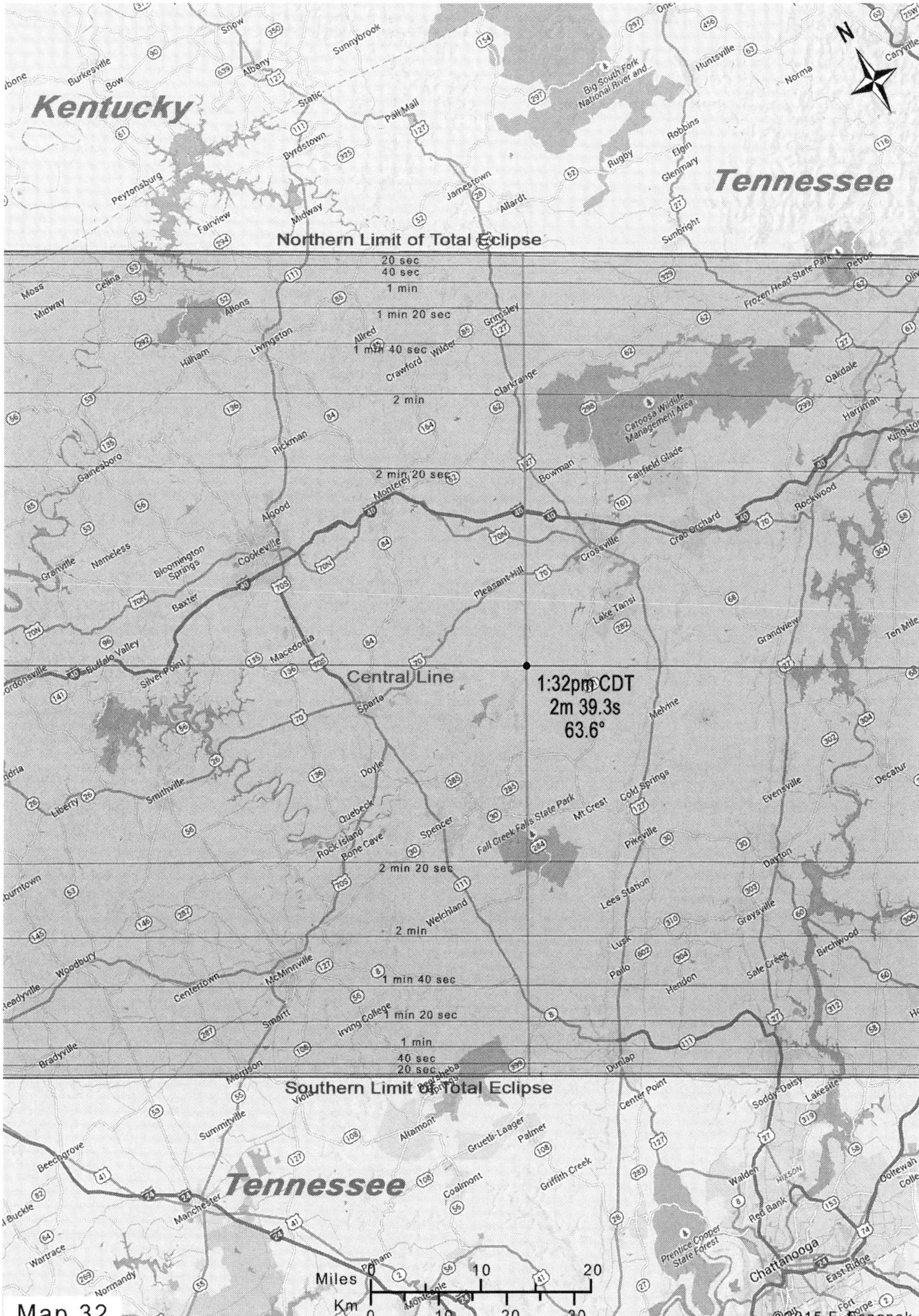

N
Kentucky
Tennessee
Northern Limit of Total Eclipse
20 sec
40 sec
1 min
1 min 20 sec
1 min 40 sec
2 min
2 min 20 sec
Central Line
1:32pm CDT
2m 39.3s
63.6°
2 min 20 sec
2 min
1 min 40 sec
1 min 20 sec
1 min
40 sec
20 sec
Southern Limit of Total Eclipse
Tennessee
Burkesville
Albany
Sunnybrook
Pall Mall
Jamestown
Allardt
Rugby
Robbins
Elgin
Glenmary
Sunbright
Huntsville
Norma
Big South Fork National River and...
Peytonsburg
Byrdstown
Static
Fairview
Midway
Moss
Celina
Hilham
Livingston
Allred
Crawford
Wilder
Grimsley
Clarkrange
Catoosa Wildlife Management Area
Frozen Head State Park
Petros
Oliver
Oakdale
Harriman
Kingston
Rickman
Gainesboro
Monterey
Bowman
Fairfield Glade
Rockwood
Crab Orchard
Crossville
Algood
Cookeville
Granville
Nameless
Bloomington Springs
Baxter
Pleasant Hill
Lake Tansi
Grandview
Ten Mile
Buffalo Valley
Silver Point
Macedonia
Sparta
Melvine
Smithville
Liberty
Doyle
Quebeck
Rock Island
Bone Cave
Spencer
Fall Creek Falls State Park
Mt Crest
Cold Springs
Pikeville
Evensville
Decatur
Dayton
Lees Station
Welchland
Graysville
Birchwood
Woodbury
Centertown
McMinnville
Lusk
Pailo
Hendon
Sale Creek
Irving College
Smartt
Bradyville
Morrison
Dunlap
Center Point
Soddy-Daisy
Lakesite
Summitville
Viola
Altamont
Gruetli-Laager
Palmer
Beechgrove
Coalmont
Griffith Creek
Manchester
Walden
Red Bank
Ooltewah
Prentice Cooper State Forest
Chattanooga
East Ridge
Wartrace
Normandy
Pelham
Monteagle
Miles
0
10
20
Km
0
10
20
30
Map 32
Map data ©2015 Google
©2015 F. Espenak

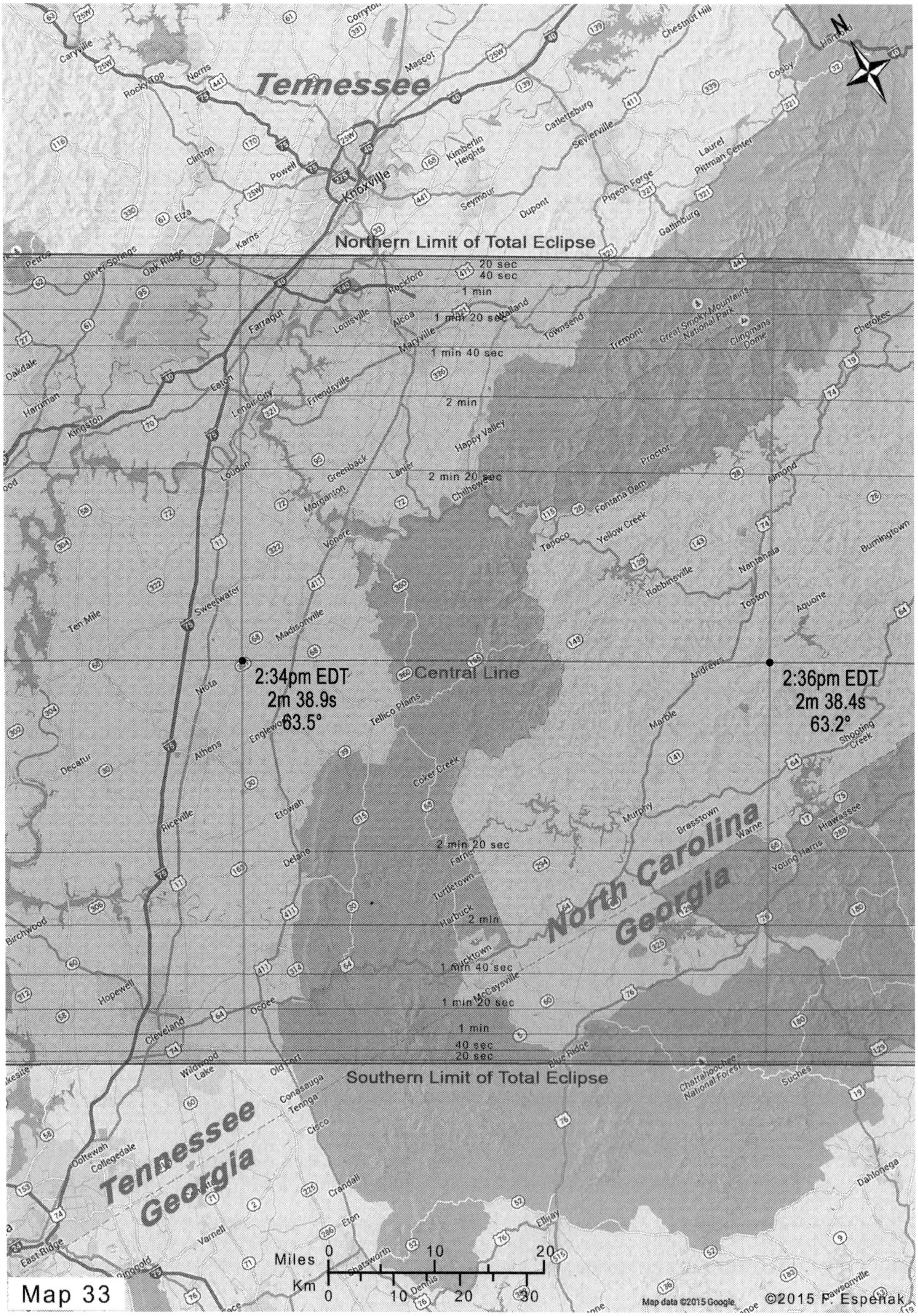
Tennessee
Knoxville
Northern Limit of Total Eclipse
20 sec
40 sec
1 min
1 min 20 sec
1 min 40 sec
2 min
2 min 20 sec
Great Smoky Mountains National Park
Clingmans Dome
Central Line
2:34pm EDT
2m 38.9s
63.5°
2:36pm EDT
2m 38.4s
63.2°
2 min 20 sec
2 min
1 min 40 sec
1 min 20 sec
1 min
40 sec
20 sec
Southern Limit of Total Eclipse
North Carolina
Georgia
Tennessee
Georgia
Chattahoochee National Forest
Miles 0 10 20
Km 0 10 20 30
Map 33
Map data ©2015 Google
©2015 F. Espenak

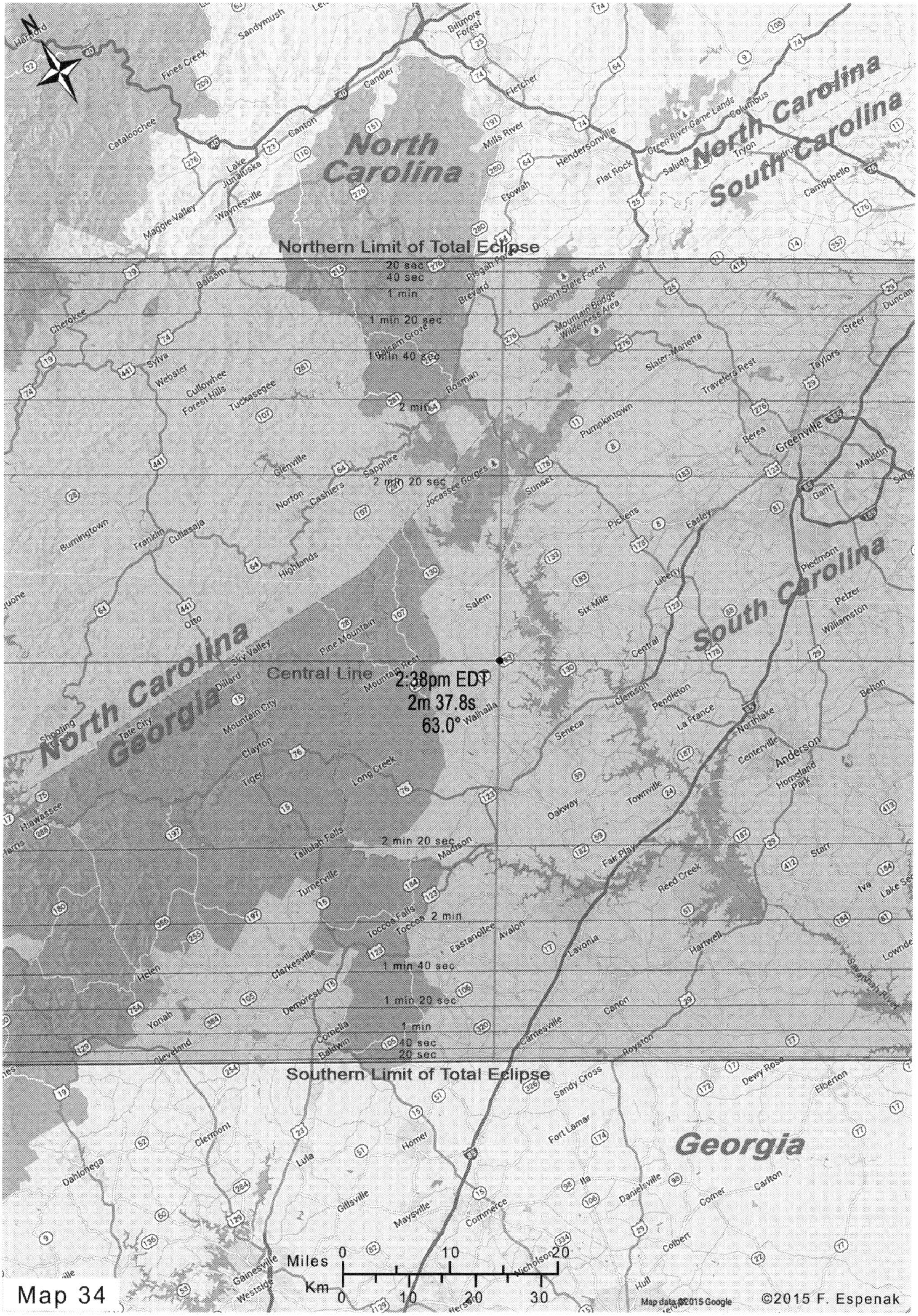

North Carolina
South Carolina
Georgia
Northern Limit of Total Eclipse
20 sec
40 sec
1 min
1 min 20 sec
1 min 40 sec
2 min
2 min 20 sec
Central Line
2:38pm EDT
2m 37.8s
63.0°
2 min 20 sec
2 min
1 min 40 sec
1 min 20 sec
1 min
40 sec
20 sec
Southern Limit of Total Eclipse
Asheville
Greenville
Anderson
Clemson
Seneca
Walhalla
Toccoa
Brevard
Hendersonville
Gainesville
Miles 0 10 20
Km 0 10 20 30
Map 34
Map data ©2015 Google
©2015 F. Espenak

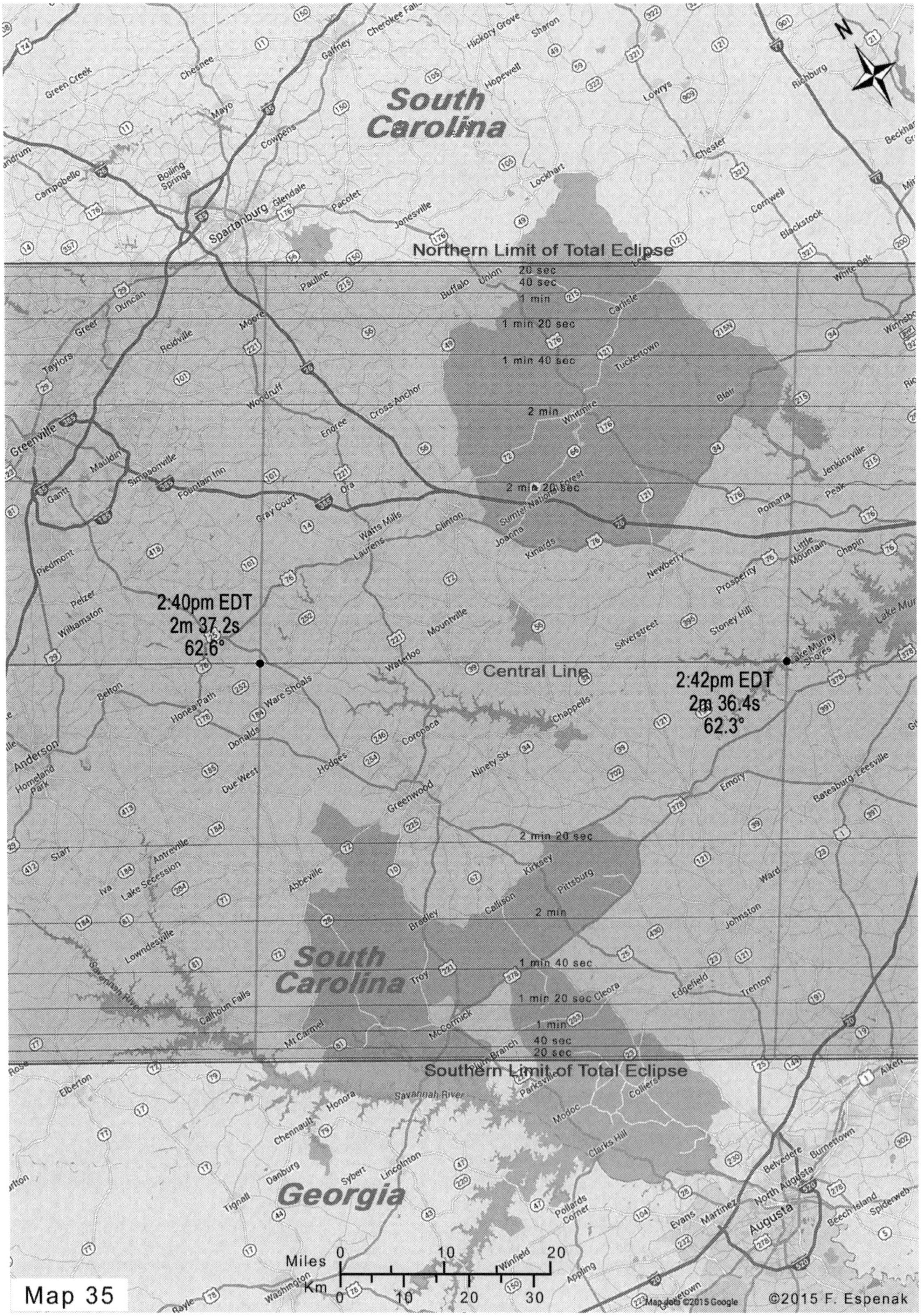
South Carolina
Northern Limit of Total Eclipse
20 sec
40 sec
1 min
1 min 20 sec
1 min 40 sec
2 min
2 min 20 sec
2:40pm EDT
2m 37.2s
62.6°
Central Line
2:42pm EDT
2m 36.4s
62.3°
2 min 20 sec
2 min
1 min 40 sec
1 min 20 sec
1 min
40 sec
20 sec
Southern Limit of Total Eclipse
South Carolina
Georgia
Spartanburg
Greenville
Anderson
Greenwood
Newberry
Clinton
Laurens
Abbeville
Augusta
North Augusta
Savannah River
Miles 0 10 20
Km 0 10 20 30
Map 35
Map data ©2015 Google
©2015 F. Espenak

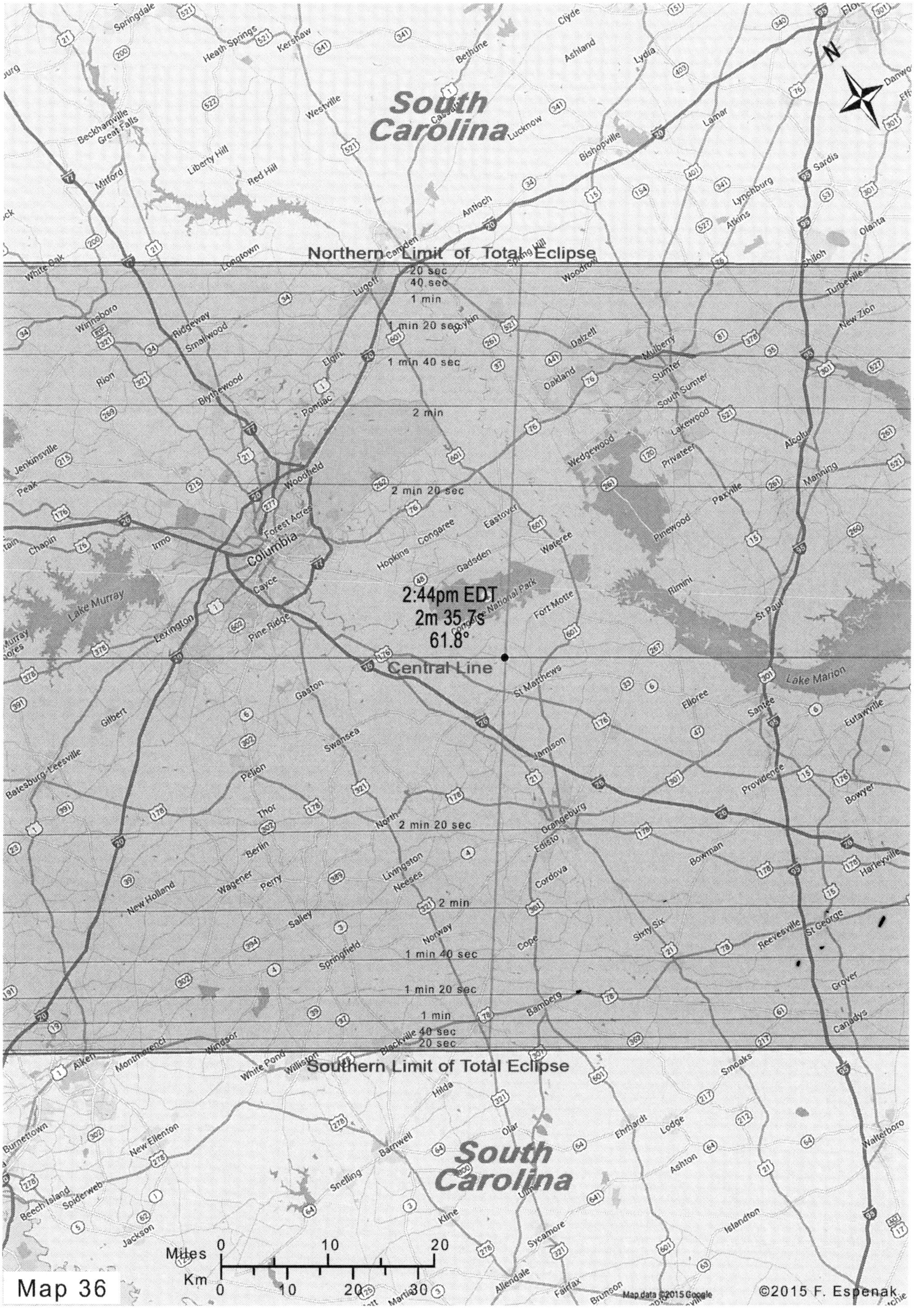
South Carolina
Northern Limit of Total Eclipse
20 sec
40 sec
1 min
1 min 20 sec
1 min 40 sec
2 min
2 min 20 sec
2:44pm EDT
2m 35.7s
61.8°
Central Line
2 min 20 sec
2 min
1 min 40 sec
1 min 20 sec
1 min
40 sec
20 sec
Southern Limit of Total Eclipse
South Carolina
N
Springdale
Heath Springs
Kershaw
Bethune
Clyde
Ashland
Lydia
Westville
Cassatt
Lucknow
Lamar
Bishopville
Beckhamville
Great Falls
Liberty Hill
Red Hill
Mitford
Antioch
Lynchburg
Atkins
Sardis
Olanta
Shiloh
Turbeville
New Zion
White Oak
Longtown
Camden
Woodrow
Lugoff
Winnsboro
Ridgeway
Smallwood
Boykin
Dalzell
Mulberry
Sumter
South Sumter
Elgin
Oakland
Rion
Blythewood
Pontiac
Lakewood
Privateer
Wedgewood
Alcolu
Manning
Jenkinsville
Peak
Woodfield
Forest Acres
Paxville
Pinewood
Chapin
Irmo
Columbia
Hopkins
Congaree
Eastover
Wateree
Gadsden
Congaree National Park
Lake Murray
Cayce
Lexington
Pine Ridge
Fort Motte
Rimini
St Paul
Lake Marion
St Matthews
Elloree
Santee
Eutawville
Gaston
Gilbert
Swansea
Jamison
Batesburg-Leesville
Pelion
Providence
Bowyer
Thor
North
Orangeburg
Edisto
Berlin
Bowman
Harleyville
Wagener
Perry
Livingston
Neeses
Cordova
New Holland
Salley
Norway
Sixty Six
Reevesville
St George
Springfield
Cope
Grover
Bamberg
Canadys
Aiken
Montmorenci
Windsor
White Pond
Williston
Blackville
Smoaks
Hilda
Burnettown
New Ellenton
Barnwell
Olar
Ehrhardt
Lodge
Ashton
Walterboro
Snelling
Ulmer
Beech Island
Spiderweb
Jackson
Kline
Sycamore
Islandton
Allendale
Fairfax
Brunson
Miles
0
10
20
Km
0
10
20
30
Map 36
Map data ©2015 Google
©2015 F. Espenak

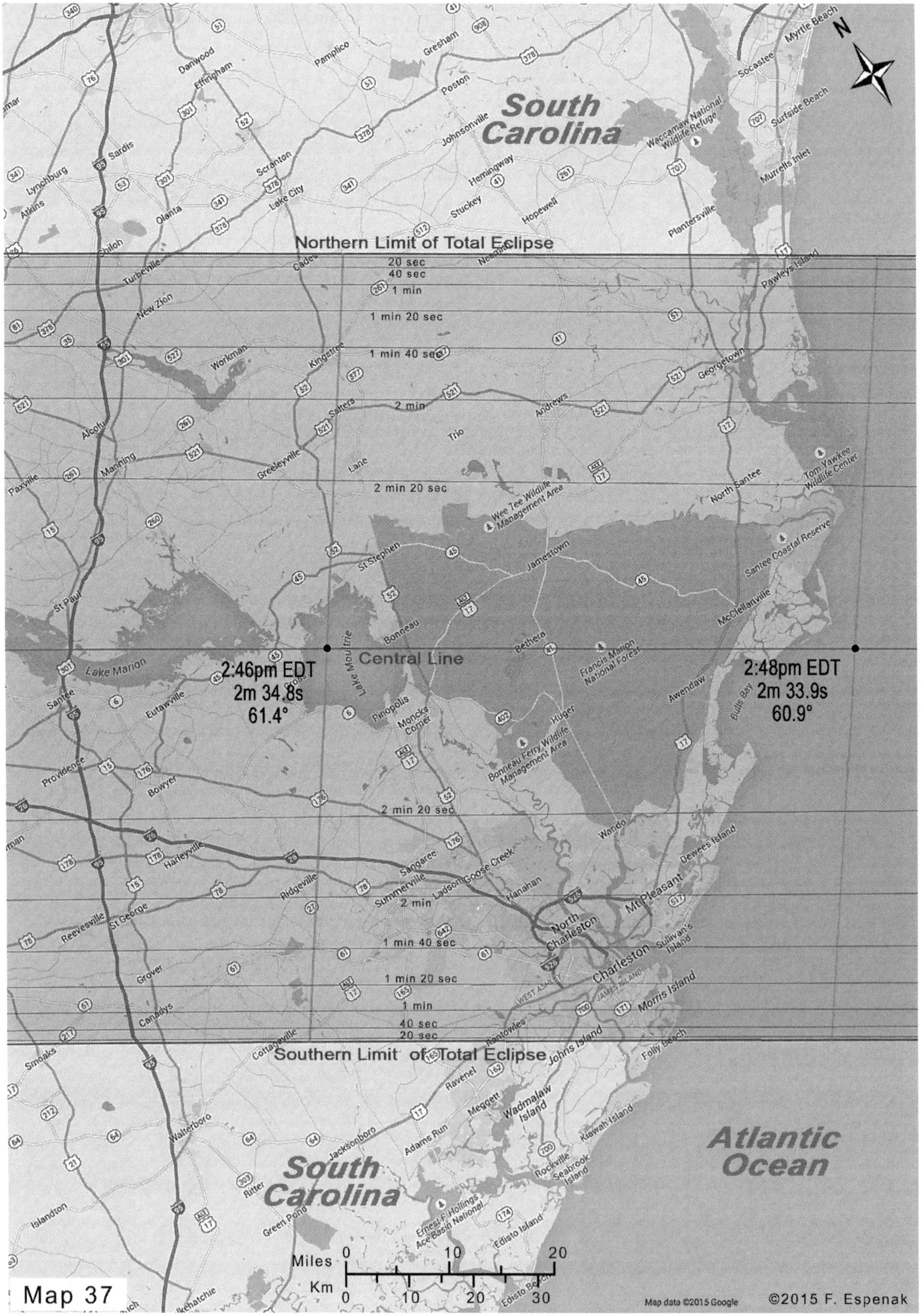
South Carolina
Northern Limit of Total Eclipse
20 sec
40 sec
1 min
1 min 20 sec
1 min 40 sec
2 min
2 min 20 sec
Central Line
2:46pm EDT
2m 34.8s
61.4°
2:48pm EDT
2m 33.9s
60.9°
2 min 20 sec
2 min
1 min 40 sec
1 min 20 sec
1 min
40 sec
20 sec
Southern Limit of Total Eclipse
Atlantic Ocean
South Carolina
Lake Marion
Lake Moultrie
Francis Marion National Forest
Wee Tee Wildlife Management Area
Bonneau Ferry Wildlife Management Area
Santee Coastal Reserve
Tom Yawkee Wildlife Center
Waccamaw National Wildlife Refuge
Ernest F. Hollings Ace Basin National
Myrtle Beach
Socastee
Surfside Beach
Murrells Inlet
Pawleys Island
Georgetown
North Santee
McClellanville
Awendaw
Bulls Bay
Mt Pleasant
Dewees Island
Sullivan's Island
Charleston
North Charleston
Morris Island
Folly Beach
Johns Island
James Island
Kiawah Island
Seabrook Island
Edisto Island
Wadmalaw Island
Summerville
Moncks Corner
Pinopolis
Bonneau
St Stephen
Jamestown
Kingstree
Lake City
Manning
Santee
Eutawville
Walterboro
Miles
0
10
20
Km
0
10
20
30
Map 37
Map data ©2015 Google
©2015 F. Espenak

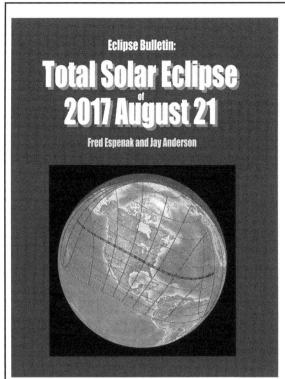

35710548R00030

Made in the USA
San Bernardino, CA
02 July 2016